Lecture Notes in Mathematics

Edited by A. Dold and B. Eckmann

617

Keith J. Devlin

The Axiom of Constructibility:

A Guide for the Mathematician

Springer-Verlag
Berlin Heidelberg New York 1977

Author
Keith J. Devlin
Department of Mathematics
University of Lancaster
Lancaster/England

Library of Congress Cataloging in Publication Data

Devlin, Keith J
The axiom of constructibility.

(Lecture notes in mathematics ; 617)
Bibliography: p.
Includes index.
1. Axiom of constructibility. I. Title. II. Series: Lecture notes in mathematics (Berlin) ; 617.
QA3.L28 no. 617 [QA248] 510'.8s [511'.3] 77-17119

AMS Subject Classifications (1970): 02K15, 02K25, 02K05, 04-01, 04A30, 20A10, 20K35, 54D15

ISBN 3-540-08520-3 Springer-Verlag Berlin Heidelberg New York
ISBN 0-387-08520-3 Springer-Verlag New York Heidelberg Berlin

Printed in Germany

Printing and binding: Beltz Offsetdruck, Hemsbach/Bergstr.
2140/3140-543210

PREFACE

Consider the following four theorems of pure mathematics.

The Hahn-Banach Theorem of Analysis: If F is a bounded linear functional defined on a subspace $\underset{\sim}{M}$ of a Banach space $\underset{\sim}{B}$, there is an extension of F to a linear functional G on $\underset{\sim}{B}$ such that $\|G\| = \|F\|$.

The Nielsen-Schreier Theorem of Group Theory: If $\underset{\sim}{G}$ is a free group and $\underset{\sim}{H}$ is a subgroup of $\underset{\sim}{G}$, then $\underset{\sim}{H}$ is a free group.

The Tychonoff Product Theorem of General Topology: The product of any family of compact topological spaces is compact.

The Zermelo Well-Ordering Theorem of Set Theory: Every set can be well-ordered.

The above theorems have two things in common. Firstly they are all fundamental results in contemporary mathematics. Secondly, none of them can be proved without the aid of some powerful set theoretical assumption: in this case the Axiom of Choice.

Now, there is nothing wrong about assuming the Axiom of Choice. But let us be sure about one thing: we are making an assumption here. We are saying, in effect, that when we speak of "set theory", the Axiom of Choice is one of the basic properties of sets which we intend to use. This is a perfectly reasonable assumption to make, as most pure mathematicians would agree. Moreover (and here we are at a distinct advantage over those who first advocated the use of the Axiom of Choice), we know for sure that assuming the Axiom of Choice does not lead to a contradiction with our other (more fundamental) assumptions about sets.

In Chapter I of this book we describe four classic open problems of mathematics, as above one from Analysis, one from Algebra, one from General Topology, and one from Set Theory. Since we call these "problems" rather than "theorems", however, it should be obvious that they are not quite the same as our four statements above. Indeed, it can be shown that assuming the axiom of choice does not lead to a solution of any of these problems. But by making a further assumption about sets, we are able to solve each of these problems (and many more problems known to be unsolvable without such an assumption). This assumption is the Axiom of Constructibility.

The Axiom of Constructibility is an axiom of set theory. It is a natural axiom, closely bound up with what we mean by "set". It implies the axiom of choice. It is known not to contradict the more basic assumptions about sets. And as we have already indicated, its assumption leads to the solution of many problems known to be unsolvable from the Axiom of Choice alone. Time alone will tell whether or not this axiom is eventually accepted as a basic assumption in mathematics. Currently, the situation is not unlike that involving the Axiom of Choice some sixty years ago. The axiom is being applied more and more, and what is more it tends to decide problems in the "correct" direction. And one can provide persuasive arguments which justify the adoption of the axiom. (Again as with the Axiom of Choice in the past, there are also arguments against its adoption.) However, since the axiom is being applied in different areas of pure mathematics, it is a proposition of interest to the mathematician at large regardless of the final outcome concerning its "validity".

Until recently the notion of a constructible set was studied extensively only by the mathematical logician. Indeed, any kind of in-depth study requires a considerable acquaintance with the ideas and methods of mathematical logic — in particular, the notions of formal languages, satisfaction, model theory, and a good deal of pure set theory. But with the growing use of the Axiom of Constructibility in areas outside of set theory, the axiom has become of interest to mathematicians who do not possess all of these prerequisites from logic. It is for this audience that we have written this short account. Our basic premise in writing has been that, whilst it would be very nice if everyone had at least a basic knowledge of elementary mathematical logic, this is almost certainly not the case. We therefore assume no prior knowledge of mathematical logic. (The one exception is Chapter V, but the book is designed so that this chapter can be totally ignored without affecting anything else.) Since it would clearly be far too great a task to develop this material to a level adequate for anything approaching a comprehensive treatment of constructibility, we choose instead to cut some corners and arrive at the required definitions very quickly. In other words, we present here a <u>description</u> of set theory and the Axiom of Constructibility, not the theory itself. Admittedly this approach may prove annoying to logicians — but they do not need to read this account, being well equipped to consult a more mathematical account.

The book is divided up as follows. In Chapter I we discuss some well known problems of pure mathematics. Since each of these problems is unsolvable on the basis of the current system of set theory, but can be solved if one assumes the Axiom of Constructibility, they provide both a motivation for considering the axiom, as well as illustrations of its application. In Chapter II we give a brief account of set theory. This forms the basis of our description of constructibility in Chapter III. Chapter IV applies the Axiom of Constructibility in order to solve the problems considered in Chapter I. Chapter V is different from the rest of this book in that some knowledge of logic is assumed.(At least, for a full appreciation of our discussion a prior knowledge of logic is required. The reader may be able to gain some idea of what is going on without such knowledge. We certainly try to keep things as simple as possible.) In Chapter V we try to explain just <u>how</u> it is that the Axiom of Constructibility enables one to answer questions of mathematics of the kind considered in the previous chapters. In order to illustrate our description we present a further application of the axiom, this time in Measure Theory. (We thereby provide some consolation for measure theorists who may have felt left out by our choice of problems in Chapter I.) The book is structured on the assumption that many readers will not wish to go into the subject matter of Chapter V very thoroughly, if at all.

It is to be hoped that mathematicians may wish to use the Axiom of Constructibility. For this reason the proofs in Chapter IV are given in some detail, except that in each case we state without proof a very general combinatorial principle which is a consequence of the Axiom of Constructibility, and then use this principle in order to prove the desired result. The advantage of this approach is that the reader may use the proof as a model for other proofs, without having to spend a great deal of time investigating the Axiom of Constructibility itself.

Finally a word about our use of the phrase "pure mathematics". In writing this book it has been convenient to restrict the meaning of this phrase to "pure mathematics other than set theory". A similar remark applies to our use of the word "mathematician" in our title.

CONTENTS

Chapter I FOUR FAMOUS PROBLEMS

In order both to motivate the consideration of the Axiom of Constructibility, and to illustrate its use, we give here a brief account of four well known problems of pure mathematics, one from analysis, one from algebra, one from general topology, and one from set theory. These problems all have one thing in common: they cannot be solved on the basis of the usual set theoretical assumptions (axioms), but they are solvable if we assume the Axiom of Constructibility.

1. A Problem in Real Analysis[1]

Let X be an infinite set, $<$ a linear ordering of X. We may define a topology on X by taking as an open basis all intervals $(a,b) = \{ x \in X \mid a < x < b \}$ for $a,b \in X$ with $a < b$. A classic theorem of Cantor says that if X has no largest member and no smallest member, and if the above topology on X is both connected and separable, then X is (considered as an ordered topological space) homeomorphic to the real line, $\mathbb{R}$, (considered as an ordered topological space). The basic idea behind the proof is to take a countable dense subset of X (by separability), prove that this set is isomorphic to the rationals, $\mathbb{Q}$, and then show that X must be isomorphic to the Dedekind completion of the dense subset, and hence isomorphic to $\mathbb{R}$, the Dedekind completion of $\mathbb{Q}$. Use is made of the fact that the connectedness of X is equivalent to the two facts (a) that for each pair a,b of elements of X with $a < b$ there is a third element, c, of X with $a < c < b$, and (b) that every subset of X which is bounded above/below in X has a least upper bound/ greatest lower bound in X.

It is not unnatural to ask if the above characterisation of the real line is the best possible. Can we, for instance, weaken any of the conditions on X (and $<$) and still obtain the conclusion that such an X will be isomorphic to $\mathbb{R}$? From this standpoint, the following question is quite natural. Let us say that a linearly

1. Strictly speaking, this is not a problem of real analysis itself. But it is clearly of interest to any real analyst. (We make no apology for any ambiguity in this last sentence.)

ordered set X satisfies the countable chain condition (c.c.c.) if every collection of pairwise disjoint open intervals is countable. (The reader should not worry about where "chains" get into the act. There are good historical reasons for using the word "chain" here, as well as some, not so overwhelming, mathematical reasons.) Clearly, if X is separable, then X will satisfy the c.c.c. So it is not unreasonable to pose the following question:

Let X be an infinite set, $<$ a linear ordering of X. Suppose that under the ordering $<$, X has no largest member and no smallest member. Regard X as a topological space as above. If X is connected and satisfies the c.c.c., does it follow that $X \cong \mathbb{R}$?

This question was first raised by M. Souslin in 1920. It soon became known as The Souslin Problem. Curiously enough, although the question is so basic and so very simple to pose, it resisted numerous attempts at solution over the next forty years. Of course (?), in view of the important role played by the fact that the reals have a countable dense subset, one would expect the Souslin Problem to have a negative answer. But no counterexample was forthcoming. We shall see why in the ensuing chapters.

2. A Problem in Algebra

We consider now a famous problem of group theory. As a first step, let us establish the convention that "group" will always mean "abelian group".

Let G, A, B be groups. We say that G is an extension of A by B iff A is a subgroup of G (written $A \triangleleft G$) and $G/A \cong B$. (Thus B describes, in a sense, the manner in which G extends A.)

Given groups G and A with G an extension of A, there is a unique (up to isomorphism) group B such that G is an extension of A by B: namely the group G/A. The extension problem (for abelian groups) asks the following converse question. Given groups A and B, determine the extensions of A by B. There is always at least one such: namely, the direct sum $A \oplus B$. But there may be more than one. For instance, let $\underset{\sim}{Z}$ be the group of integers, $2\underset{\sim}{Z}$ the subgroup of the even integers, and $\underset{\sim}{2}$ the unique group of order 2. Now, $\underset{\sim}{Z}/2\underset{\sim}{Z} \cong \underset{\sim}{2}$, so $\underset{\sim}{Z}$ is an extension of $2\underset{\sim}{Z}$ by $\underset{\sim}{2}$. But $\underset{\sim}{Z} \not\cong 2\underset{\sim}{Z} \oplus \underset{\sim}{2}$ since $\underset{\sim}{Z}$ is torsion free and $2\underset{\sim}{Z} \oplus \underset{\sim}{2}$ has torsion. The following solution to the extension problem is due to Baer.

Let G, G′ be extensions of A by B. Thus $G/A \cong B$, $G'/A \cong B$. Let

$$G \xrightarrow{\varphi} B \qquad \text{and} \qquad G' \xrightarrow{\varphi'} B$$

be the canonical projections. We write $G \sim G'$ iff there is an isomorphism

$$G \xrightarrow{\psi} G'$$

which makes the following diagram commute:

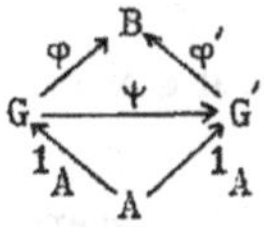

(As usual, 1_A denotes the identity morphism on A into whatever extension of A is being considered.)

It is easily seen that $\sim$ is an equivalence relation on the set of all extensions of A by B. The relation $G \sim G'$ is stronger than isomorphism, and is clearly the correct notion of "sameness" when we are considering extensions of A by B.

R. Baer proved, in 1949, that the set of equivalence classes of extensions of A by B under the above equivalence relation itself forms a group. We denote this group by Ext(B,A). It is the group of extensions of A by B. It is outside the scope of this book to describe the group operation involved in Ext(B,A), and indeed we shall not need to know it. What we do need to know is that the identity element of the group Ext(B,A) is the equivalence class of the direct sum $A \oplus B$.

We should also mention that as a result of work by Schreier it is possible to give a description of the members of Ext(B,A) in terms of A and B.

Let us recall now that a group G will be free iff there is a set $\{g_i \mid i \in I\}$ of elements of G such that every non-zero element of G has a unique representation of the form

$$n_1 g_{i_1} + \cdots + n_k g_{i_k} ,$$

where $i_1,\ldots,i_k$ are distinct members of I and $n_1,\ldots,n_k$ are non-zero integers.

We then say $\{g_i \mid i \in I\}$ is a basis for G. A basis is thus the same as a linearly independent generating set. We relate the notion of a free group to the extension problem as follows.

2.1 Theorem

Let $H \triangleleft G$. If G/H is free, then $G = H \oplus N$ for some $N \triangleleft G$, $N \cong G/H$.

Proof: Let $\{H + k \mid k \in K\}$ be a basis for G/H, where $K \subseteq G$. Let $N = \langle K \rangle_G$, the subgroup of G generated by K. It is easily checked that $G = H \oplus N$. □

2.2 Corollary

If B is free, then $\mathrm{Ext}(B,A) = 0$.

Proof: Let $A \triangleleft G$, $G/A \cong B$. Since G/A is free, 2.1 gives $G \cong A \oplus B$. (More precisely, the proof of 2.1 shows that $G \sim A \oplus B$ in the sense defined above.) □

The above result has a converse. In order to obtain the converse, we recall the following standard theorem.

2.3 Theorem (Nielsen-Schreier)

If G is free and $H \triangleleft G$, then H is free. □

2.4 Lemma

Let B be a given group. If $\mathrm{Ext}(B,A) = 0$ for all groups A, then B is free.

Proof: Let F be the free group on B (i.e. the unique group which is freely generated by the set B). Let

$$F \xrightarrow{\varphi} B$$

extend the identity function on B. Set $A = \mathrm{Ker}(\varphi)$. Then F is an extension of A by B, so by hypothesis, $F \cong A \oplus B$. Hence there is an embedding

$$B \longrightarrow F.$$

So by 2.3, B is free. □

Now , by 2.2, if G is free, then $\mathrm{Ext}(G,\mathbb{Z}) = 0$. J. H. C. Whitehead asked, in 1951, if this statement has a valid converse. In other words, does the property that $\mathrm{Ext}(G,\mathbb{Z}) = 0$ characterise the free groups G ? Defining a W-group to be any group G for which $\mathrm{Ext}(G,\mathbb{Z}) = 0$, this reduces to showing that every W-group is free. Until recently, the only result of note on this problem was the following, proved in 1951:

2.5 Theorem (Stein)

Every countable W-group is free. □

We shall return to the Whitehead Problem in later chapters.

3. A Problem in General Topology

Let $(X,\mathfrak{T})$ be a topological space. The following separation properties which X may satisfy are well known.

X is T_0 if, whenever $x,y \in X$ and $x \neq y$ there is an open set U which contains exactly one of x, y.

X is T_1 if, whenever $x,y \in X$ and $x \neq y$ there are open sets U, V such that $x \in U$, $y \notin U$ and $x \notin V$, $y \in V$.

X is T_2 (Hausdorff) if, whenever $x,y \in X$ and $x \neq y$ there are disjoint open sets U, V such that $x \in U$ and $y \in V$.

X is regular if, whenever $x \in X$ and $A \subseteq X$ is closed and $x \notin A$, there are disjoint open sets U, V such that $x \in U$ and $A \subseteq V$. X is T_3 if it is regular and T_1.

X is normal if, whenever A, B are disjoint closed subsets of X there are disjoint open sets U, V such that $A \subseteq U$ and $B \subseteq V$. X is T_4 if it is normal and T_1.

It is immediate that $T_4 \rightarrow T_3 \rightarrow T_2 \rightarrow T_1 \rightarrow T_0$. None of these implications is reversible.

The following generalisation of the Hausdorff property occurs in the literature. We say a subset Y of X is discrete if every point of X has a neighborhood which contains at most one point of Y. (Thus if X is T_1, a subset Y of X will be discrete iff Y has no limit points in X, which says that the elements of Y are spaced well apart from each other.) A space is collectionwise Hausdorff if, whenever Y is a discrete set, there is a family $\{U_y \mid y \in Y\}$ of pairwise disjoint open sets such that $y \in U_y$ for all $y \in Y$. (We call such a family a separation of Y.)

Now, it is not hard to show that the Hausdorff property does not in general imply the collectionwise Hausdorff property. Indeed, there are T_4 spaces which are not collectionwise Hausdorff. So we ask what extra conditions on a space are required in order to yield the conclusion that it be collectionwise Hausdorff ? Considerations outside of our present scope lead to the following precise question (which is fairly close to the best possible). Recall that a space is said to satisfy the first axiom of countability if for each point x the neighborhood system of x has a countable basis. The question now is: Is every first countable T_4 space collectionwise Hausdorff? We investigate this problem in IV.4. Let us finish by mentioning that this question first arose out of research on a very famous problem of General Topology — the normal Moore space problem (which is still open). This problem deals with the metrization problem (i.e. which topological spaces are metric spaces ?). Roughly speaking, what the normal Moore space problem asks is whether every first countable T_4 space is collectionwise normal.

4. A Problem in Set Theory

The oldest of our four problems — the continuum problem — dates back to Cantor. The question raised here is: How many real numbers are there ? In order to make this precise we require some elementary notions from set theory.

Fundamental in mathematics is the notion of counting. And it is to be expected that our reader is familiar with (at least some of!) the natural numbers 0,1,2,... . Using the natural numbers we may "count" the "number" of elements in any finite set. But what about infinite sets ? Well, why not extend the natural number system into the transfinite ? Why not indeed! By doing this we obtain the ordinal number system, which commences with the natural numbers, and which is adequate to "count" the elements of <u>any</u> set. What <u>are</u> the ordinal numbers ? We answer this question by first answering the question: What <u>are</u> the natural numbers ?

The number 0 we <u>define</u> to be the empty set, $\emptyset$. The number 1 we <u>define</u> to be the set $\{0\}$ (i.e. the set with precisely one element, that element being the natural number 0). The number 2 we <u>define</u> to be the set $\{0,1\}$. Proceeding inductively, we <u>define</u> the number n+1 to be the set $\{0,1, \ldots ,n\}$. Notice that the number n is always a set with exactly n elements, those elements being precisely the numbers smaller than n. To obtain the ordinal numbers we continue the definition into the transfinite. The first infinite ordinal, denoted by ω, is the set

$$\{0,1, \ldots ,n, \ldots \quad \ldots \ldots\}$$

of all natural numbers. The second infinite ordinal, $\omega+1$, is the set

$$\{0,1, \ldots ,n, \ldots \quad \ldots \ldots, \omega\}.$$

In general, the next ordinal number after α will be the set $\alpha \cup \{\alpha\}$. And when we have defined the sequence of ordinals

$$0,1, \ldots ,n, \ldots ,\omega,\omega+1, \ldots ,\alpha, \ldots \qquad ,$$

this sequence having no last member, the "next" ordinal number will be the set

$$\{0,1, \ldots ,n, \ldots ,\omega,\omega+1, \ldots ,\alpha, \ldots\}$$

of all ordinals constructed so far.

In general we use lower case Greek letters to denote ordinal numbers. Notice that by our definition of ordinal number, if α, β are ordinal numbers, α will precede β (i.e. be <u>smaller</u> than β), written $\alpha < \beta$, just in case $\alpha \in \beta$. Thus the ordinal numbers are totally ordered by $\in$. Indeed they are well-ordered by $\in$. Moreover, regarded as the set of all smaller ordinal numbers, each ordinal number is itself well-ordered by $\in$.

Now, each ordinal number has associated with it a canonical well-ordering : namely $\in$. It can be shown that every well-ordered set P can be put into an order-preserving, one-one correspondence with a unique ordinal, called the order-type of P, written otp(P). In this way the ordinal numbers can be used to "count" the number of elements in any well-ordered set (otp(P) being the answer for the well-ordered set P). But by Zermelo's Well-Ordering Theorem (which is a consequence of the Axiom of Choice), every set can be well-ordered. Hence we may use the ordinal numbers to "count" the elements of any set. The problem here is that the result of our counting depends upon the well-ordering chosen. Now in the case of finite sets we are used to the fact that it does not matter in which order we count the elements of that set ; the answer will always be the same. But for infinite sets this is no longer the case. Different well-orderings of the same set can lead to different results to the process of counting the elements of that set. For example, consider the set, ω, of natural numbers. Under the usual well-ordering, this set, ω, has order-type ω . But we can also well-order the set ω as follows:

$$\{0,2,4, \ldots \ldots, 1,3,5, \ldots \ldots\} .$$

Under this well-ordering the order-type of the same set ω is the ordinal number $\omega+\omega$, which is the second ordinal constructed by taking the set of all previous ordinals when that set has no largest member. Thus, although we can use the ordinal numbers for "counting" infinite sets, they are really only suited for measuring the size of well-ordered sets. Fortunately, however, using the ordinal numbers we may obtain a number system which is able to "count" the elements of an arbitrary set. We make use of the fact that the ordinal numbers are themselves well-ordered by $\in$.

Given a set X, we define its cardinality to be the least ordinal number α which may be put into one-one correspondence with X. Equivalently, the cardinality of X is the least order-type of all well-orderings of X. The cardinality of X is usually denoted by $|X|$. Any ordinal number which equals $|X|$ for some set X is called a cardinal number. It is immediate from this definition that an ordinal number α will be a cardinal number iff it cannot be put into one-one correspondence with any smaller ordinal number. Clearly, $0,1,2, \ldots, n, \ldots$ are all cardinal numbers. So too is ω. But $\omega+1$, $\omega+2$, etc. are not cardinal numbers, since each may be put into one-one correspondence with ω . The first cardinal after ω is denoted by ω_1, the next by ω_2, and so on. Thus for any ordinal number α, ω_α denotes the αth cardinal after ω. We usually reserve κ, λ, μ to denote cardinals. Also, since we sometimes have occasion to consider a cardinal number both as a cardinal and as an ordinal, it is convenient to have a notation which indicates the usage. On occasions where the distinction is important we use the Hebrew letter "aleph", with $\aleph_\alpha$ denoting ω_α considered as a cardinal. (We write $\aleph_0$ instead of ω.)

With the cardinal numbers we now have a good system for measuring the size of a set X. Every set X will have associated with it a unique cardinal number, its cardinality. A set X will be countable iff its cardinality is at most ω (i.e. at most $\aleph_0$). Now, a classical result of <u>Cantor</u> tells us that the set of all real numbers, $\mathbb{R}$, is uncountable. Hence $|\mathbb{R}| \geqslant \aleph_1$. We know that there is a unique ordinal α such that $|\mathbb{R}| = \aleph_\alpha$. So what is this α ? Cantor himself conjectured that the α here was 1, i.e. that $|\mathbb{R}| = \aleph_1$. But the problem resisted all attempts at solution over the years: it is known as the <u>continuum problem</u>.

It is possible to rephrase the continuum problem slightly. If κ is a cardinal, then $\kappa = \{\alpha \mid \alpha < \kappa\}$, so we may consider the set of all functions from κ into 2, denoting this set by ${}^{\kappa}2$. We define the cardinal 2^κ to be $|{}^{\kappa}2|$. (More generally, we define cardinal exponentiation by setting $\kappa^\lambda = |{}^{\lambda}\kappa|$.) By considering binary representations, we see that $|\mathbb{R}| = 2^{\aleph_0}$. So the continuum problem asks for the α which solves the equation $2^{\aleph_0} = \aleph_\alpha$. The <u>continuum hypothesis</u>, CH, is the assertion $2^{\aleph_0} = \aleph_1$. The <u>generalised continuum hypothesis</u>, GCH, is the statement that for all α, $2^{\aleph_\alpha} = \aleph_{\alpha+1}$. Denoting the next cardinal after κ by κ^+, we may write the GCH in the alternative form $(\forall\kappa)(2^\kappa = \kappa^+)$, it being understood from the notation here that κ ranges over all infinite cardinals.

We shall return later to the continuum problem. In the meantime, whilst we are considering ordinal numbers, let us introduce one or two further notions which we shall require. Firstly, note that our ordinal numbers fall into two categories. If α is an ordinal number, then $\alpha = \{0,1,2, \ldots, \gamma, \ldots\} = \{\gamma \mid \gamma < \alpha\}$.

Now, either it is the case that α has a largest member,

$$\alpha = \{0,1,2, \ldots ,\gamma, \ldots ,\beta\} \ ,$$

or else α has no largest member,

$$\alpha = \{0,1,2, \ldots ,\gamma, \ldots\} \ .$$

In the former case, α is the first ordinal number after β (i.e. $\alpha = \beta+1$), and we say α is a <u>successor ordinal</u>. In the latter case, for every ordinal $\beta < \alpha$ there is an ordinal strictly between β and α (in particular, $\beta+1$ is such an ordinal), and we say α is a <u>limit ordinal</u>. For example, all positive integers are successor ordinals, whilst ω is a limit ordinal. It is easily seen that, in fact, every infinite cardinal number is a limit ordinal.

For readers totally unfamiliar with ordinal numbers, let us make the following observations. The ordinal numbers are an intuitive continuation of the natural numbers into the transfinite. As we proceed upwards we meet, in turn, a copy of the natural numbers, then a limit ordinal, then another copy of the natural numbers, then another limit ordinal, and so on, <u>ad infinitum</u>. Unfortunately, the presence of this last phrase <u>ad infinitum</u> results in this picture being at the same time both accurate and misleading. The point is, there will be many limit ordinals which do not at all have the characer of ω , being points at which a collection of much greater magnitude takes place. We touch on this point in IV.1, but unfortunately it would constitute far too great a digression to be precise here. At any rate, the picture we have just painted seems better than no picture at all.

We may carry out definitions and proofs by induction on the ordinal numbers just as with the natural numbers, except that we now have the new case of limit ordinals to deal with. For example, we may define addition of ordinal numbers by means of the following induction (induction on β for arbitrary α):

$\alpha + 0 = \alpha$; $\alpha + (\beta+1) = (\alpha + \beta) + 1$; $\alpha + \beta = \bigcup_{\gamma<\beta} (\alpha+\gamma)$, if β is a limit ordinal. (There is some ambiguity of notation here, since the symbol ' +1 ' is used to denote 'next ordinal', but the very definition of ordinal addition makes this difference unimportant.) Further details can be found in any elementary text.

Chapter II WHAT IS SET THEORY ?

1. Set Theory as a Framework for Mathematics

The use of elementary set theoretical notions and reasoning is, of course, prevalent in all branches of pure mathematics, and needs no elaboration here. But for the most part, scant attention is paid to the set theory itself. For instance, although a group is defined to be a set together with an operation (i.e. function) defined on that set, the precise nature of that set is never considered. This is, of course, because it is totally irrelevant (in group theory) what the set actually consists of: what matters is the behaviour of the group operation defined on the set. In other words, the structure is what counts, not the material used to support that structure. By and large the same is true for any part of algebra. In analysis the situation is a little different, since it is not irrelevant to ask the question: "What are the real numbers?" It is really a matter of opinion as to whether the construction of the real line belongs to the field of analysis, of course, since a quite common (and entirely reasonable) approach is to take the real number system as basic. Another common approach is to take as basic, say, the integers, and then to construct the rationals, and thence the real numbers as Dedekind cuts. Or, taking this process one step further we may commence with just the natural numbers. But this is about as far as any analyst would bother to go. But there is no real reason to stop here. As we saw in I.4, the natural numbers can themselves be defined as certain sets. Indeed, in order to construct the natural numbers we need only make one basic existence assumption: namely that nothing exists! So, we see that it is possible to construct the entire real number system starting with nothing. The assumption that the natural numbers exist has been dispensed with. Thus, not only does set theory provide us with a useful framework for discussing structures, be they algebraic or analytic, it also answers any relevant questions of existence (modulo the assumption that nothing exists).

The above discussion would seem to require some elaboration. The idea that we may construct even the natural numbers by assuming only the existence of nothing will doubtless strike the reader as a little suspicious. Well, how did we define the natural numbers? We assumed the existence of the empty set (i.e. nothing), and took this to be the number 0. We then took for the number 1 the set whose only member was this set 0. Having 0 and 1 available, we were then able to define 2 as the set whose elements were precisely 0 and 1. And so on. In the end we obtained all natural numbers. But wait. Whilst it is certainly true that the only raw existence assumption was that there is a set with no elements, we did allow ourselves one basic operation: given any collection of objects (sets), we may regard that entire collection as an object (set) in its own right. This is what set theory is all about. We reduce existence problems to operations on sets. Given the one basic operation that any given collection of sets is itself a set, and starting only from the existence of the empty set, we may construct all of the sets required in mathematics. So much (for now) for existence. What about the various other concepts of mathematics ? For instance, what about functions? Well, the definition (or interpretation, if you prefer) of a function as a set of ordered pairs is well known. And we may define the ordered pair of two objects (sets) x and y by

$$(x,y) = \{\{x\},\{x,y\}\}.$$

(It is easily verified that $(x,y) = (x',y')$ iff $x = x'$ & $y = y'$, so this definition is perfectly acceptable.) Thus the ordered pair operation may be defined set-theoretically. Hence we obtain functions as particular sets. (In fact, we may without loss of generality assume that all the functions of mathematics are unary! Indeed, from a set-theoretical standpoint they are, because the n-tuple $(x_1,\dots,x_n)$ is defined inductively by $(x_1,\dots,x_n) = ((x_1,\dots,x_{n-1}),x_n)$, and is thus nothing more than a single set, and an n-ary function is just a function which is only defined on sets which are n-tuples.) Besides the function concept, however, we may define all the other concepts of mathematics within set theory. Thus, not only does set theory provide us with the existence of the sets we require, it also provides us with the various operations and notions which we need.

To finish this section, let us point out that our above discussion is not intended to convey anything particularly deep or profound. Indeed, if taken on that kind of level it is possible to raise all sorts of questions concerning our development. But, hopefully, we managed to convey the general idea of how set theory provides us with a unified, fundamental framework for mathematics, so that any branch of (pure) mathematics can be regarded as set theory restricted to particular sets. (Lest the overenthusiastic reader be carried away by this last statement, maybe we should say that in most instances there is absolutely no point in actually regarding a branch of mathematics in this way, so it will still be necessary to read other books in mathematics than books on set theory. Indeed, it very often comes as a great suprise to all concerned when a problem in, say, algebra, actually does require a set-theoretical approach for its solution.)

2. Set Theory Under the Microscope

Having seen how set theory provides us with a unified framework for all of pure mathematics, it is now high time that we examine this rather powerful set concept. Since everything in mathematics is now a set, just what is a set ? We have said that, besides assuming the existence of the empty set, there is only one basic notion in set theory:

> The ability to regard any collection of objects as an object in its own right.

O.K., fine. But what determines a collection ? In the case of finite collections there is (in theory) no problem: we may determine a finite collection by listing precisely what the elements of the set are to be. But in the case of arbitrary collections this is no longer possible. The only possibility is to give some property which characterises the elements of the collection concerned. Thus, if P is some property, the "collection" of all those x for which P(x) holds is a well-defined collection. As usual, we shall denote by

$$\{ x \mid P(x) \}$$

the collection of all x for which P(x) is true.

But the above discussion has only postponed the day of reckoning. No sooner have we defined what we mean by "collection" than we are faced with the question: What is a property ? A simple answer would be to allow any "property" which can be described in the English language. But this is much too vague, since it will include such "collections" as the set of all bicycles or the set of all mathematicians , neither of which should be included in a mathematical theory of sets. Indeed, our definition should only include collections of genuine mathematical objects. On the other hand, we do not want to exclude any bona fide mathematical collections. So, a not unreasonable approach would be to define a language which will describe those and only those collections we require for mathematics. The reader who has not seen anything of this nature before may well be suprised to discover that this is indeed possible. He will be even more suprised at the simplicity of the language.

3. A Language for Set Theory

The notion of a formal language is familiar to all computer users. Programming languages are such. That is, they have a fixed (and highly restricted) vocabulary, and fixed (and very rigid) rules of grammar. Nevertheless, despite the apparent inflexibility of a programming language, it is well known that it is usually adequate for expressing highly complex procedures. So it is with the language of set theory we define below.

We commence by listing the basic symbols (words) of our language.

Variables : v_n $(n = 0,1,2,\ldots)$;

Logical connectives : $\wedge$, $\vee$, $\neg$, $\rightarrow$, $\leftrightarrow$;

Quantifier symbols : $\forall$, $\exists$;

Brackets : (,) ;

Set-theoretic symbols : $\in$, $=$.

The variables are to be used to denote the objects of our discussion, i.e. sets. Since we do not wish to place any prior restriction on the number of sets we may at any one time discuss, we allow ourselves an infinite collection of variables, even though we shall never be (theoretically) able to use them all.

The logical connectives have the following meanings: $\wedge$ denotes <u>and</u> , $\vee$ denotes <u>or</u> (the non-exclusive variety), $\neg$ denotes <u>not</u>, $\rightarrow$ denotes <u>implies</u> , and $\leftrightarrow$ denotes <u>iff</u>. Had we wished, we could of course have used these words themselves rather than introduce symbols for them, and this would have made the analogy with programming languages much clearer, but on the whole it seems better to adopt the usual practice and use the symbols. (For one thing, sentences are actually easier to "read" when the symbols are used.)

The quantifier symbols read: for $\forall$ read <u>for all</u> , and for $\exists$ read <u>there exists</u> . . . <u>such that</u> .

The brackets serve as punctuation symbols, denoting the start and finish of a "clause" or "sentence".

Finally we have the two symbols which make the language a set-theoretic language. We want our language to be adequate for describing any mathematical concept. Taking it as known that we may describe any such concept in set-theoretic terms, we see that what is required is that our language is strong enough to describe any concept of set theory. Suprisingly enough, it suffices that, besides the "standard apparatus" already described, all we require is that our language be capable of expressing just two fundamental notions:

x <u>is an element of</u> y

and x <u>is equal to</u> y.

We introduce the symbol $\in$ to denote <u>is an element of</u> and $=$ to denote <u>is equal to</u>.

That then is our language. It is called the language of set theory, or LST for short. Despite its rudimentary nature, it is powerful enough to express all the concepts of mathematics which we use. Before we can indicate how this may be "proved", we must of course say what the rules of grammar are to be.

As basic formulas (or basic clauses) we allow all expressions of the form

$$(v_n \in v_m) \quad , \quad (v_n = v_m) \quad .$$

We then build up the formulas (i.e. the clauses and sentences) by applying the following rules:

if φ , ψ are formulas, so are $(\varphi \wedge \psi)$, $(\varphi \vee \psi)$, $(\neg \varphi)$, $(\varphi \to \psi)$, $(\varphi \leftrightarrow \psi)$.

if φ is a formula, then so are $(\forall v_n \varphi)$, $(\exists v_n \varphi)$.

The meanings of the above rules should be quite clear in view of our previous discussion.

Now, the reader unfamiliar with this kind of thing may well have trouble believing that this language is capable of expressing all the concepts we need in mathematics. Nevertheless, this is true. Of course, for all but the most basic of concepts, the formula of LST which defines the concept will be incredibly complex, and totally unreadable. (Just as a typical computer program is virtually unreadable.) But we are not interested in actually expressing concepts in LST. What is important is that it is always possible. And how do we prove this? Well, we clearly cannot give a proof in any rigorous sense, since, on the one hand the collection of all properties expressible in LST is a precisely defined collection, whereas the collection of all concepts we use in mathematics is a vague, undefined collection. So what we do is as follows. We first develop an extensive software. For readers unfamiliar with computing, let us elaborate this point by saying that what we do is use our language in order to successively define more and more complex notions. Once a new notion has been defined, we thence treat it as a regular part of the language. In a remarkably speedy manner, this process leads us to the usual "language" of mathematics (by which we mean the "language" which mathematicians use when writing up proofs). At which point we can safely say that our task has been accomplished.

In order to illustrate the above outline, let us see how the process we just described starts off. Let us expand LST to include some of the familiar notions of set theory. Remember, what we must do is at each stage define our new concept from concepts which are already available.

$v_1 \subseteq v_2$: $(\forall v_0((v_0 \in v_1) \rightarrow (v_0 \in v_2)))$

$v_1 = \cup v_2$: $(\forall v_0((v_0 \in v_1) \leftrightarrow (\exists v_3((v_0 \in v_3) \wedge (v_3 \in v_2)))))$

$v_1 = \{v_0\}$: $(\forall v_2((v_2 \in v_1) \leftrightarrow (v_2 = v_0)))$

$v_2 = \{v_0, v_1\}$: $(\forall v_3((v_3 \in v_2) \leftrightarrow ((v_3 = v_0) \vee (v_3 = v_1))))$

$v_2 = v_0 \cup v_1$: $v_2 = \cup\{v_0, v_1\}$

$v_2 = (v_0, v_1)$: $(\forall v_3((v_3 \in v_2) \leftrightarrow ((v_3 = \{v_0\}) \vee (v_3 = \{v_0, v_1\}))))$

v_0 <u>is an ordered pair</u> : $(\exists v_1(\exists v_2(v_0 = (v_1, v_2))))$

v_0 <u>is a function</u> : $(\forall v_1((v_1 \in v_0) \rightarrow (v_1$ <u>is an ordered pair</u>$))) \wedge$

$(\forall v_1 \forall v_2 \forall v_3((((v_1, v_2) \in v_0) \wedge ((v_1, v_3) \in v_0)) \rightarrow (v_2 = v_3)))$.

etc.

Hopefully, by now we have convinced the reader that LST will indeed prove adequate for our needs. We will certainly have convinced him of the pointlessness of ever actually <u>using</u> LST in order to do mathematics. So why have we introduced this powerful yet cumbersome language ? Well, the point is, since LST is precisely defined, the collection of properties describable within LST is a well defined collection. It is immediate that any property which is describable in LST will be a bona fide mathematical property. And by our above discussion, LST envelopes all of the expressive power of the everyday "language of mathematics". We shall thus be able to define a collection to be any "collection" defined by a property expressible in LST.

4. The Set-Theoretic Hierarchy

At this point it is very tempting to argue as follows.

(I) The basic premise of set theory is that given any collection of objects (sets), we may regard the entire collection as an object (set) in its own right.

(II) In order to obtain an adequate and meaningful set theory, we allow all (and only those) properties which can be expressed in LST to determine collections.

(III) By (I) and (II), given any property P which is expressible in LST, the collection $\{x \mid P(x)\}$ is a set.

The above argument certainly leads to a well-defined theory of sets. Moreover, all the sets we need in mathematics will be there. The only trouble is that the theory is inconsistent! Indeed, the inconsistency is easily derived. Consider the property $P(x) : x \notin x$. (i.e. x is not an element of x) P can be expressed in LST (the formula $(\neg(v_0 \in v_0))$ does it). So, assuming (III) above, $\{x \mid P(x)\}$ is a set. Call this set y. Since y is a set in its own right, it must be the case that either P(y) is true or that P(y) is false. But look, if P(y) holds, then y must fulfill the definition of P, so it must be the case that $y \notin y$, whence y cannot satisfy the property which defines y, whence P(y) is false. Thus, if P(y) is true then it is false. Hence it must be the case that P(y) is false. Thus y does not satisfy P, which means that it must be the case that $y \in y$, which means that y must satisfy the property which defines y, which means that P(y) is true. We have thus arrived at a contradiction.

So what has gone wrong ? The answer is that we have been careless in going from (I) and (II) to (III). We interpreted (II) correctly. (Indeed, the destructive property P considered above is so basic that there is no avoiding having this property around for the definition of sets.) But we did not interpret (I) correctly. Let us reconsider (I), emphasising the point where we went wrong. (I) says:

GIVEN any collection of sets, we may regard the entire collection as a set.

The point is, we can only form a new set out of sets which are already at our disposal. We cannot include sets which are not there! Thus the crucial word GIVEN in (I). Let us see why our ignoring this one word led to the catastrophe it did. Let P be any property expressible in LST. (e.g. Let P(x) say $x \notin x$.) We wish to consider the collection of all those x for which P(x) holds. But what are all those x ? Certainly, until we have formed the collection $\{x \mid P(x)\}$, this collection will not be available to be considered as one of the all those x. Indeed, there may be many other "sets" which will not be available until afterwards. So how do we escape from what is beginning to look like a hopeless situation? Well, the answer is already there in our basic premise (I) above. Fundamental to set theory is the notion of a hierarchy of sets. We build sets. They are not handed to us on a silver platter.

But no, already there is another possible snag. How do we build sets? Well, you say, we use the properties expressible in the language LST. If P is a property expressible in LST, then at any stage in our inductive process of building sets we may form as a new set the collection of all those x which are available which satisfy the property P. But look, what do we mean by saying that an x satisfies P ? In order to know whether x satisfies P we must check whether P(x) is true or not. But P may well contain in its definition quantifiers of the form

for all z ... , or

there is a z such that ... .

Thus, in order for the statement "P(x) is true" to be at all meaningful, must it not be the case that the universe of all sets is already a well defined collection? (For how do we know what "for all z ..." means if we do not know what all the sets are?) There are two ways out of this dilemma. We consider first the classical one of E. Zermelo.[1]

The basic idea is as follows. We take as basic the iterative building of sets approach at which we most painfully arrived above, but ignore for the moment the problem about actually defining sets by means of properties expressible in LST. This is achieved by introducing the notion of the unrestricted power set operation. If x is any set, we denote by $\mathcal{P}(x)$ the collection of all subsets of x. Thus:

$$\mathcal{P}(x) = \{ y \mid y \subseteq x \} .$$

For the time being we make no comment as to just what sets are in $\mathcal{P}(x)$. Hence our use of the word "unrestricted" above. In order to obtain our universe of sets now, we commence with the empty set, $\emptyset$, and iterate, ad infinitum, the power set operation. In order to make this more precise, we need first to introduce the ordinal number system.

Before we can have a hierarchy, we must have some means of indexing the hierarchy. For a hierarchy of sets we require the ordinal numbers. So we shall take this system as given. This can be avoided, but only at the loss of some degree of clarity. So our approach is similar to that of the analyst who takes as basic the natural numbers and proceeds from there to construct the real line. In an appendix we see how to eliminate our assumption of the ordinal number system from the development, replacing the assumption by a system of axioms from which everything can be defined. And this is akin to the elimination of the assumption of the system of natural numbers by introducing the Peano axioms. The reason why few authors commence with the Peano axioms is, of course, that it is far clearer to work with the intuitive picture of the natural numbers than with a bunch of abstract axioms.

---

1. It is the second solution which gives us the Axiom of Constructibility.

And the same is true here. The ordinal number system is just as intuitive as is the natural number system. Indeed, it simply makes precise the natural idea of "counting", only with the ordinal numbers we allow the counting to extend into the transfinite. As to the structure of the ordinal number system, the description given in I.4 will suffice. The class of all ordinal numbers will be denoted by On. The class On is not a <u>set</u> in the sense of our universe of sets, of course, just as the set of all natural numbers is not a natural number. (A better analogy would be that the set of natural numbers is not a finite set!) This is because our hierarchy of sets should continue <u>ad infinitum</u>, in the sense appropriate to set theory. (This is quite different from the "infinite" nature of the natural numbers. Although in a sense the natural numbers continue "for ever", in the sense of set theory they constitute quite a small set, and are thus fairly insignificant when compared with the universe of all sets.)

We define now the set-theoretic hierarchy. The $\alpha^{\underline{th}}$ level of this hierarchy will be denoted by V_α . (As usual, α denotes an arbitrary ordinal number.)

To commence, we start with the empty set. Thus:

$$V_0 = \emptyset .$$

Next we form all possible subsets of the sets we have so far. Thus:

$$V_1 = \{\emptyset\} .$$

Thus, although V_0 had no members, already V_1 has a member. Next we repeat the process, forming all possible sets from what is now available. This gives:

$$V_2 = \{ \emptyset, \{\emptyset\} \} .$$

Repeating the process, we obtain next:

$$V_3 = \{ \emptyset , \{\emptyset\} , \{\{\emptyset\}\} , \{\emptyset,\{\emptyset\}\} \} .$$

In general, for every positive integer n, V_n will have 2^{n-1} elements, as is readily checked. Hence V_4 is already sufficiently large to make enumerating its members a pointless exercise, so let us now write down the general rule. If V_n is defined, we set:

$$V_{n+1} = \mathcal{P}(V_n) .$$

So what do we do when we have defined V_n for every natural number n ? The answer to this question is inherent in the notion of a limit ordinal. At stage ω we just collect together everything we have so far. Thus:

$$V_\omega = \bigcup_{n<\omega} V_n .$$

Notice that, although we have not defined any new sets at stage ω , we have nevertheless made some progress. V_ω is quite a different collection from each of the previous collections V_n. Indeed, V_ω is infinite, whereas each V_n is finite. We now continue as before, setting:

$$V_{\omega+1} = \mathcal{P}(V_\omega) .$$

In general now, if V_α is defined, we set

$$V_{\alpha+1} = \mathcal{P}(V_\alpha) ,$$

and if α is a limit ordinal and we have defined V_β for all $\beta<\alpha$, then we set

$$V_\alpha = \bigcup_{\beta<\alpha} V_\beta .$$

It is not hard to see that for all α ,

$$V_\alpha \cap On = \{ \beta \mid \beta < \alpha \} = \alpha .$$

In particular,

$$V_\omega \cap On = \omega ,$$

so $V_{\omega+1}$ contains all subsets of ω . Thus, each V_n is finite, V_ω is countably infinite, and $V_{\omega+1}$ is uncountable. Thus, despite the rather slow start, once the hierarchy gets going it grows at a phenomenal rate. (Indeed, its growth is exponential, in a very precise sense.)

It is also easily seen that if $\alpha < \beta$, then $V_\alpha \subseteq V_\beta$. Thus our hierarchy is cumulative.

Now, the idea behind the above construction was that the sets of mathematics should be precisely those collections which we can construct by iterating the process of building new sets out of old ones. Thus, if we denote by V the collection of all sets, we shall require:

(P1) $V = \bigcup_{\alpha \in On} V_\alpha$.

The principle (P1) can be taken as a definition of set. We call the collection V of all sets the universe of sets.

But what has happened to our language LST in all of this? Can we now dispense with it altogether? Well no. So far we have not said very much about the power set operation. All we know is that $\mathcal{P}(x)$ is to consist of all subsets of x. Certainly, since we commence with just the empty set, V will contain nothing other than genuine, mathematical sets. But will it contain all the sets we need? In order to ensure that it does, we now place some restriction on the operation $\mathcal{P}$, in that we specify some of the sets which $\mathcal{P}(x)$ must contain. (This restriction on $\mathcal{P}$ must, of necessity, come after the event, so as to speak, as we indicated earlier, but our development is not circular.) We thus formulate our second principle.

(P2) Principle of Subset Selection. If x is a set and P is a property which is expressible in LST, then $\{ y \in x \mid P(y) \}$ is a set.

The above principle requires some comment. Firstly, the reader may wonder why we do not formulate (P2) with V_α in the place of x. Certainly we have indicated as much in our previous discussion. The point is the two versions are equivalent, so we have chosen to give the version more commonly used. Moreover, formulated as above, the principle is clearly a partial description of the power set operation.

Our second comment concerns the meaning of the phrase "expressible in LST" in (P2). By definition, if the property P involved is <u>expressible in LST</u> there must be some formula of LST which describes P. The intention is that this should enable us to define any of the sets we require in mathematics. Now, when we define sets in mathematics, we usually refer to other sets in the process. (To take a trivial example, when we define the set of all irrational real numbers we refer to the set of all real numbers.) Thus, in the description of P we should be able to refer to any sets we please. This we allow. In other words, there is no restriction placed upon the way we use the variables in the LST formula describing P in order to denote specific sets. Hence, the variables involved in this description fall into three distinct categories. Some of the variables will be used in conjunction with quantifiers, and hence are internal to the formula. Others will be used to denote specific sets. And one further variable will behave as a genuine <u>variable</u>. This will be the variable corresponding to the y in P(y): the set variable which will, according to the definition of our set, range over the set $\{ y \in x \mid P(y) \}$.

It is probably instructive to re-examine our original contradiction in the light of our new definition of sets. So let P be the property $x \notin x$. For each set x the collection $y = \{ z \in x \mid P(z) \}$ will be a set according to (P2). But now if we apply the argument which led previously to a contradiction we just end up with the positive result that $y \notin x$. And now we can say just what the collection $\{ x \mid P(x) \}$ is. For let x be any set. Then there is an ordinal α such that $x \in V_\alpha$. There is thus a least such α. Since no new sets appear at limit levels in the hierarchy, α will be of the form $\beta+1$ for some β. Then $x \subseteq V_\beta$ but $x \notin V_\beta$. Since $x \subseteq V_\beta$, we have

$$y \in x \longrightarrow y \in V_\beta .$$

Hence, as $x \notin V_\beta$, we have

$$y \in x \longrightarrow y \neq x.$$

Thus, in particular, $x \notin x$. But x was arbitrary. Hence every set satisfies the property P. In other words $\{x \mid P(x)\} = V$. Not only does the property P not lead to a contradiction, it is universally true!

That then is set theory as developed by Zermelo. In honour of Zermelo and Fraenkel, who first made the above procedure rigorous (which we have not bothered to do here), we denote this system of set theory by ZF. Thus, ZF is that system of set theory which has for its basic principles the two principles (P1) and (P2) above. (For a more accurate account of what the ZF system entails, the reader should consult Appendix I.)

5. The Axiom of Choice

With the system ZF we now have a servicable, natural set theory, which corresponds to our intuition about sets, and which is adequate for the development of mathematics from set theory. Well, almost, but not quite. Since the early part of this century the Axiom of Choice (AC, for short) has been regarded as a justifiable assumption about sets. This principle has several formulations. The most basic is the following.

Let x be a set (of sets, naturally). A choice function for x is a function f from x to sets such that for each a in x, $f(a) \in a$. (Thus f chooses an element from each member of x.) Now, in ZF one can prove that if x is finite and each element of x is non-empty, then x has a choice set. But if x is infinite this is not possible. (To be precise, we know for sure that in the system ZF one cannot prove that every set of non-empty sets has a choice function. Just how one may prove that a statement about sets is not provable in a certain system is indicated in Appendix II.) Nonetheless, although we cannot prove that choice functions always exist, it is generally thought to be a reasonable assumption that they do. So we guarantee this by introducing an additional axiom to our system. This is the Axiom of Choice, or AC for short:

AC : Every set of non-empty sets has a choice function.

AC has various other equivalent formulations. One of these is that the Cartesian product of any family of non-empty sets is non-empty. Another is Zermelo's Well-Ordering Principle: Every set can be well-ordered. Still another is the well known Zorn's Lemma: If $\underset{\sim}{P}$ is a partially ordered set such that every totally ordered subset of $\underset{\sim}{P}$ has an upper bound in $\underset{\sim}{P}$, then $\underset{\sim}{P}$ has a maximal element. (In each case the equivalence with AC is provable in the system ZF.)

We denote by ZFC the system ZF together with the Axiom of Choice. To be precise, our system ZFC is what is commonly referred to as Zermelo-Fraenkel Set Theory, but the meanings of the notations ZF and ZFC are now standard in set theory, despite the misleading aspect of these abbreviations.

Chapter III THE AXIOM OF CONSTRUCTIBILITY

Using techniques described in Appendix II, it has been proved that the four problems of pure mathematics described in Chapter I cannot be solved on the basis of the system ZFC. In view of the fundamental nature of these questions, this surely indicates a serious deficiency in our set theory. So what went "wrong" ? Well, we surely cannot abandon the principle of subset selection, (P2), since this is the one principle which is indispensible to us if we are to develop a set theory which is adequate for mathematics. So we must re-examine (P1). The reason why ZFC is not adequate for the solution of our four problems must lie in the fact that we took as a basic, undefined notion, the unrestricted power set operation, $\mathcal{P}$. So let us go back to the point where we felt it necessary to introduce this idea. Our argument then was that before we could introduce the principle of subset selection we must have at our disposal all the sets. But is this the case ? What if we stay even closer to our "iterative building of sets" idea than before ? Indeed, let us regard the original idea of sets being <u>describable collections</u> as basic. Let us define our hierarchy of sets as follows. We shall start with the empty set, and at limit levels we shall collect everything together as before. But in proceeding from stage α to stage $\alpha+1$ we shall introduce just those subsets of what is available which we can describe using LST. "But look," you say, "we had to abandon this kind of approach before, because until we have all the sets available we do not know whether any statement expressed in LST is true or not." Well, if by "true" you mean <u>really true</u> (i.e. in the universe V), then this is correct. But consider the following alternative. At stage α we have constructed some sets. This collection of sets can be regarded as a <u>partial universe</u>. (Indeed, it <u>is</u> such!) And relative to this partial universe, any statement expressible in LST (which refers explicitly only to sets in this partial universe) will either be true or false. Hence, if at stage $\alpha+1$ we say to ourselves "We shall build all sets we can using LST properties, <u>referring only to what is available right now</u> , " then we can indeed proceed without trouble. In other words, we shall construct our new hierarchy almost as if we were actually <u>carrying out</u> the construction step by step, at each stage having only those sets available which we have constructed so far, and introducing new sets only when we can construct them using properties expressible in LST. This reasoning leads us to the definition of the constructible hierarchy.

1. The Constructible Hierarchy

By analogy with the Zermelo hierarchy, V_α, $\alpha \in On$, we define the constructible hierarchy of sets, L_α, $\alpha \in On$. We shall commence as before with the empty set, and at limit stages we shall just collect everything together that we have so far, again as before. But in passing from L_α to $L_{\alpha+1}$ we shall not take "all" subsets of L_α (whatever that means), but only those collections from L_α which are, in a precise sense to be outlined below, describable in terms of L_α.

Suppose φ is some formula of LST. Then φ will in general contain, amongst other things, quantifiers of the form $\forall v_n$ or $\exists v_n$. And in general, the variable v_n will range over all sets. More precisely, the intended domain of the quantifier is V. Suppose now that M is some set which contains all of the sets which any variables in φ may denote. By φ^M we shall mean the formula φ with all quantifiers restricted to M. Of course, in a strict sense φ^M is exactly the same formula as φ. What is different is the way we interpret φ^M and φ. A quantifier $\forall v_n$ in φ is read as "for all sets v_n", whereas a quantifier $\forall v_n$ in φ^M is read as "for all sets v_n in M". Notice that there need be no connection between the validity of φ and the validity of φ^M. For instance (and in terms of ZF set theory), let φ be the sentence

$$\exists v_0(v_0 \text{ is an uncountable cardinal})$$

(expressed in LST, of course). Now, ω_1 is the first uncountable cardinal, and

$$V_{\omega_1} \cap On = \omega_1 = \{\alpha \mid \alpha < \omega_1\},$$

so $\varphi^{V_{\omega_1}}$ is false. But since $\omega_1 \in V_{\omega_2}$, $\varphi^{V_{\omega_2}}$ is true, as is φ itself. The point is that in $\varphi^{V_{\omega_1}}$, the quantifier $\exists v_0$ only ranges over V_{ω_1}, which contains no uncountable cardinal. (The reader may be disturbed by the fact that this example is constructed in ZF, a system which we appear to have abandoned. He should contain his concern for a while, for we have a pleasant suprise in store for him.)

Given any set M now, we denote by Def(M) the collection of all subsets of M which can be defined by a formula of the form φ^M, where φ is a formula of LST such that M contains any sets which variables in φ denote. We call Def(M) the family of all definable subsets of M. Notice that since the formulas φ^M involved in this definition only refer (in any manner) to members of M, we only require the availability of M in order to define Def(M).

We define the constructible hierarchy as follows:

$$L_0 = \emptyset \quad ;$$

$$L_{\alpha+1} = \mathrm{Def}(L_\alpha) \quad ;$$

$$L_\alpha = \bigcup_{\beta<\alpha} L_\beta \text{ , if } \alpha \text{ is a limit ordinal} \quad .$$

Comparing this with the Zermelo hierarchy, we clearly have $L_\alpha \subseteq V_\alpha$ for all α. In fact, since we may explicitly define any finite set (by enumerating its elements), we have $L_n = V_n$ for all natural numbers n, whence also $L_\omega = V_\omega$. But $L_{\omega+1} \neq V_{\omega+1}$, since $V_{\omega+1}$ is uncountable whereas $L_{\omega+1}$ will be countable. (These remarks presuppose that we are working in the system ZF. Once again we ask the reader to supress for a while any unease he may feel about this.)

It is easily seen that $\alpha \leqslant \beta$ implies $L_\alpha \subseteq L_\beta$, and that for any ordinal α, $L_\alpha \cap \mathrm{On} = \{\beta \mid \beta < \alpha\} = \alpha$. Thus the constructible hierarchy grows in a manner which resembles the growth of the Zermelo hierarchy. But the rate of growth of the constructible hierarchy is quite slow. Indeed, for any infinite ordinal α we shall have $|L_\alpha| = |\alpha|$. (This is because LST has only countably many formulas. Readers familiar with cardinal arithmetic should have no difficulty in supplying a proof of this equality.)

For notational convenience later on, let us define

$$L = \bigcup_{\alpha\in\mathrm{On}} L_\alpha \, .$$

We call the collection L the constructible universe (of sets).

2. The Axiom of Constructibility

The axiom of constructibility is the assumption that the collection L contains all sets. Or, to put it another way, when we assume the axiom of constructibility we define a set to be an element of L. Of course, the way we have introduced the axiom of constructibility suggests that the use of the word "axiom" here is unnecessary. But as we shall see in a moment, our terminology is in fact quite appropriate. The same remark applies to our next definition.

We have already introduced the symbol V to denote the universe of sets. And L denotes the universe of constructible sets. So we have on hand a convenient way of denoting the axiom of constructibility, namely:

$$V = L \quad .$$

Thus, what $V = L$ says is that in order to obtain all sets we start off with the empty set and successively form new sets from old by means of the Def operator. So if we work in constructible set theory, instead of taking as our basic assumptions the two axioms:

(P1) $V = \bigcup_{\alpha \in On} V_\alpha$;

(P2) Principle of subset selection ,

we take the two assumptions:

(L1) $V = \bigcup_{\alpha \in On} L_\alpha$ (i.e. $V = L$) ;

(L2) Principle of subset selection.

Let us assume now the definition of set which constructible set theory gives us. This provides us with a well-defined universe of sets. Now, within this universe for each set x there will surely be a set whose elements are precisely the subsets of x . (More on this point in a moment.) Hence, within our universe there will be a well-defined power set operator. There is thus nothing to prevent us from carrying out our previous definition of the Zermelo hierarchy, V_α ($\alpha \in On$). (Only now we are working within a fixed universe of sets, of course.) And what do we find ? Well, we find that $\bigcup_{\alpha \in On} V_\alpha = V$. In fact, for all α, $L_\alpha \subseteq V_\alpha$. (Equality can only occur very infrequently.) In other words, the principle (P1) is still valid! The only difference between what (P1) meant before and what it now means lies in the fact that whereas the power set operator is unrestricted in ZF set theory, in constructible set theory it is governed by a much more precisely defined universe of stes. Thus, constructible set theory is not an <u>alternative</u> to ZF set theory, it is an <u>extension</u> of it. In other words, we may regard the principle V = L as an axiom of set theory which we assume <u>in addition to</u> the ZF axioms. This illustrates both our use of the word "axiom" and our abbreviation for the axiom itself.

At this point we feel we owe the reader some comments concerning the assumption made above that in constructible set theory, each set has a well-defined power set. We were not claiming that this could be <u>proved</u> from the axiom of constructibility. But it is a very reasonable assumption to make when developing any theory of sets. And in the light of our comments above concerning the nature of the axiom of constructibility, we can demonstrate that it is no worse to assume it in conjunction with constructible set theory than with ZF set theory, where it plays a fundamental role. We did not explicitly list the axiom of power set existence as a basic principle of ZF set theory, although it certainly is so, because the principle (P1) already conveys the essential structure of the system. But what is the situation when we append the system ZF by the additional assumption V = L ? Could this not invalidate the axiom of power sets (and thus lead to an inconsistent theory)? The answer is no. Indeed, assuming the axiom of constructibility as an additional axiom of set theory on top of the principles of ZF is totally harmless, in view of the following classical result of Gödel:

2.1 <u>Theorem</u>

If the system ZF + V = L is inconsistent, then so is the system ZF. □

Thus, the axiom of constructibility is there for us to assume if we want to. It cannot lead us into any more trouble than could already the ZF system. (Note also from the statement of 2.1 the convenient notation we now have for constructible set theory: ZF + V = L .) At this point, readers whose pet subjects require the Axiom of Choice may be asking themselves if they could live with ZF + V = L instead of ZFC. They do not need to worry. They may safely use AC in conjunction with ZF + V = L. Indeed, in constructible set theory we do not need to assume AC as an axiom: it is a theorem:

2.2 Theorem (in the system ZF + V = L)

Every set of non-empty sets has a choice function. □

2.2 was proved by Gödel. Proofs of both 2.1 and 2.2 may be found in the more rigorous development of constructibility theory presented in our article [De] .

As the basic set theory underlying mathematics it is now customary to take the system ZFC , Zermelo-Fraenkel set theory. Whenever a result is proved using the additional assumption V = L, the statement of the result must then be prefaced by the sentence: "Assume V = L" . (This is what is customarily done when a result is proved using the GCH. Instead of saying: "In the system of set theory ZFC + GCH ....", one assumes ZFC as basic and just writes "Assuming GCH ...".

To sum up then, we arrived at the notion of constructibility by trying to obtain a good definition of set. Our approach was that a collection can only be said to exist if there is some way of describing it. (This being made precise in a fairly unrestrictive manner, allowance being made for the reference to any other sets in describing a collection.) Having obtained the notion of the constructible hierarchy, and having defined the axiom of constructibility, we observed that our previous, ZF intuition was still sound, and that all that had in effect happened was that we had removed the weakness inherent in ZF that the power set operation is taken as almost totally undescribed. Since most theorems in mathematics can be proved without any great use being made of the fundamental nature of what constitutes a set (except for recourse to AC now and then), we decided to retain the "simpler" system ZFC as basic, but we could now call on the axiom of constructibility to help us out in those cases where ZFC is not enough. Thus, if a theorem can only be proved using V = L, then we know that at the heart of this lies a fundamental question about the nature of sets. This approach does not preclude our adopting the system ZF + V = L as basic (thereby retaining the intuition of our original development). It would just be the case that for the most part it doesn't matter one way or the other (whence any approach will suffice). But for the four problems introduced in Chapter I it matters a great deal what our set theory is. This is the subject matter of the rest of this chapter and of the next.

3. The Generalised Continuum Hypothesis.

Recall that GCH is the assertion $\forall\kappa(2^{\kappa} = \kappa^{+})$. As we mentioned in I.4, the GCH resisted all attempts at proof over the years. Indeed, even the special case of the CH ($2^{\aleph_0} = \aleph_1$) went unsolved. Then in 1963, P. J. Cohen gave a rigorous proof of the fact that neither the CH nor the GCH could be settled on the basis of the system ZFC alone. (In Appendix II we give some indication of how such a result may be obtained.) This highlighted an old result of Gödel:

3.1 Theorem

Assume V = L. Then GCH is true. □

It is not possible to give a rigorous proof of this result here. (Such may be found in [De] .) But let us see what is involved. Consider the case of CH. Now, this reduces to proving that there are at most $\aleph_1$ subsets of ω . (There are clearly as many subsets of ω as there are characteristic functions of such, whence the two sets $\mathcal{P}(\omega)$ and ${}^{\omega}2$ have the same cardinality.) Consider now the construction of the constructible hierarchy. At stage $\omega+1$ some subsets of ω will appear. Perhaps by making reference to some of these we can describe more subsets of ω to come in at stage $\omega+2$. Then we may be able to describe still more sets to go into $L_{\omega+3}$, and so on. Thus, unlike the situation with the Zermelo hierarchy where, by definition, all subsets of ω appear together in $V_{\omega+1}$, in the constructible hierarchy new subsets of ω may keep appearing, as our facility for describing new collections increases. But, and this is a very fundamental fact, it can be shown that this process stops at stage ω_1. In other words, if we assume $V = L$, then all subsets of ω will appear in the first ω_1 levels of the constructible hierarchy. Thus $\mathcal{P}(\omega) \subseteq L_{\omega_1}$. But we have mentioned earlier that it is quite easy to prove that for any infinite ordinal α, $|L_\alpha| = |\alpha|$. Thus we conclude that $2^{\aleph_0} = |{}^{\omega}2| = |\mathcal{P}(\omega)| \leq |L_{\omega_1}| = \aleph_1$, proving CH. A similar argument proves GCH assuming $V = L$.

In V.4 we say a little more about the above, in that we indicate why no new sets of natural numbers occur after stage ω_1.

Thus, by assuming $V = L$ we may solve one of the four problems of Chapter I. In the next chapter we consider the three remaining problems.

4. Historical Remarks

The definition of the constructible hierarchy is due to K. Gödel, and first appeared in his classic monograph [Gö]. In his monograph, Gödel also formulated the axiom of constructibility, $V = L$. But it was not Gödel's intention that the axiom of constructibility should be taken as a basic axiom of set theory (and thence of mathematics). Rather he used constructibility as a tool in order to establish both the non-refutability in ZF of AC and the non-refutability in ZFC of GCH. (2.1 and 2.2 taken together give the first of these, and when we add on 3.1 as well we get the second.) Gödel's approach to set theory was that there is, in some sense, a "fixed" universe of sets which the Zermelo-Fraenkel axioms attempt to describe. If we adopt this approach, then there is of course no *a priori* reason why we should adopt $V = L$ as an *axiom*. If, on the other hand, we think of sets as being something that we *build* in an iterative manner, then we may certainly insist that we proceed in the manner which led us to the axiom of constructibility. (Of course, whichever course we take, we are going to end up with a system of axioms and proceed to develop our set theory rigorously from these axioms. And when we do this we shall see that in order to construct either the Zermelo hierarchy or the constructible hierarchy we must use quite a few of our axioms. But this ultimate process of standing our analysis on its head, so as to speak, does not have any bearing on the intuition which leads to our final choice.)

Had it not been for some remarkable developments over the last ten years, the Gödel approach of regarding V = L as a useful tool to establish non-refutability results would undoubtedly have remained unchallenged. For until then applications had been to statements like AC and GCH, which are highly set theoretical, and which one might expect to depend heavily on the set theory involved. But in 1967, R. B. Jensen proved from the axiom of constructibility a result of pure mathematics (as opposed to abstract set theory — see the comment on page (iv) .), known to be unprovable in Zermelo-Fraenkel set theory. Subsequent work of Jensen led to a whole chain of results about constructibility, and opened the way for various people to apply the axiom of constructibility to solve problems in different areas of pure mathematics. This stimulation led the present author at least to reconsider the status of V = L as an axiom of set theory, and to the writing of this book in particular. Of course, only time will tell the outcome (just as was the case with AC some sixty years ago). But since the axiom of constructibility has applications in several areas of mathematics now, it certainly merits attention.

Chapter IV APPLICATIONS OF $V = L$ IN MATHEMATICS

It is to be expected that proofs involving the axiom of constructibility have a quite different flavour from proofs in ZFC. Indeed, we only need to use $V = L$ in the first place because the "usual" methods fail. The same is true of proofs involving AC. A good illustration is the standard proof of the Hahn-Banach theorem of analysis. The proof falls into two distinct parts: one part being clearly of an analytic nature (more accurately, it has the flavour of Banach space theory), the other part, the application of Zorn's lemma, being set theoretical in nature. Of course, it does not take the pure mathematician long to master the use of AC. But with $V = L$ we are already at a much deeper level, and many applications of $V = L$ can really only be fully understood with a solid background in set theory and (parts of) mathematical logic. Fortunately this is not always the case, however. In his investigation of constructibility, Jensen established a collection of purely combinatorial consequences of $V = L$, and for many problems a straightforward application of one of these consequences is all that is required.

1. Combinatorial Principles from $V = L$.

Before we can state the combinatorial principles we require some notions from set theory.

The infinite cardinals fall under two categories, the singular cardinals and the regular cardinals. These are defined as follows. If α is a limit ordinal, a subset A of α is unbounded in α if there is no $\beta < \alpha$ with $A \subseteq \beta$ (equivalently, for each $\beta < \alpha$ there is a $\gamma \in A$ with $\gamma > \beta$). A function f from an ordinal δ into α is cofinal if f is order preserving and the range of f is unbounded in α. The cofinality of α, $cf(\alpha)$, is the least δ such that there is a cofinal map $f: \delta \to \alpha$. It is immediate that $cf(\alpha)$ is a limit ordinal, and indeed is an infinite cardinal. For example, $cf(\omega) = \omega$, $cf(\omega + \omega) = \omega$, $cf(\omega_\omega) = \omega$. (For this last example take $f: \omega \to \omega_\omega$

defined by $f(n) = \omega_n$.) An infinite cardinal κ is <u>regular</u> if $cf(\kappa) = \kappa$. Otherwise (in which case $cf(\kappa) < \kappa$), κ is <u>singular</u>.

It can be shown that every <u>successor cardinal</u> (i.e. a cardinal of the form κ^+ for some cardinal κ) is regular. Thus, $\aleph_1$, $\aleph_2$, ... are all regular, and so are $\aleph_{\omega+1}$ and $\aleph_{\omega_1+1}$. The <u>limit cardinals</u> (i.e. cardinals which are not successor cardinals) $\aleph_\omega$ and $\aleph_{\omega_1}$ are singular. $\aleph_0$ is a regular limit cardinal. It cannot be proved that there are any other regular limit cardinals, though it is most unlikely that this could be disproved. (Assuming V = L certainly would not help to disprove it.) Hence, any limit cardinal other than $\aleph_0$ which we meet is likely to be singular. Another definition of singularity is in terms of cardinal addition. An infinite cardinal κ is singular iff there is a cardinal $\lambda < \kappa$ and cardinals $\kappa_\alpha < \kappa$ $(\alpha < \lambda)$ such that $\kappa = \sum_{\alpha<\lambda} \kappa_\alpha$, in which case $cf(\kappa)$ is the least λ for which such $\kappa_\alpha < \kappa$ can be found. For instance, $\aleph_\omega = \sum_{n<\omega} \aleph_n$.

Let κ be any uncountable regular cardinal. A set $C \subseteq \kappa$ is <u>closed</u> if, whenever $\delta < \kappa$ is a limit ordinal and $f:\delta \longrightarrow C$ is order preserving, then $\sup_{\alpha<\delta} f(\alpha) \in C$. (Since the ordinals are well-ordered, every set of ordinals has a supremum of course.) The closed unbounded subsets of κ generate a κ-complete filter in $\mathcal{P}(\kappa)$. Indeed, if $C_\alpha \subseteq \kappa$ are closed and unbounded in κ for all $\alpha < \lambda$, where $\lambda < \kappa$, then $\bigcap_{\alpha<\lambda} C_\alpha$ is also closed and unbounded in κ. We indicate the proof for the case $\lambda = 2$, and leave the reader to generalise this to any $\lambda < \kappa$.

We are given closed and unbounded subsets A and B of the uncountable regular cardinal κ. We wish to prove that $A \cap B$ is closed and unbounded in κ. We verify closure first. Let $f:\delta \longrightarrow A \cap B$, where $\delta < \kappa$ is a limit ordinal and f is order preserving. Then $\sup f[\delta] \in A$ and $\sup f[\delta] \in B$ so $\sup f[\delta] \in A \cap B$. We check unboundedness now. Let $\alpha < \kappa$ be given. We must find a $\beta \in A \cap B$, $\beta > \alpha$. We define a function $f:\omega \longrightarrow \kappa$ by induction. Since A is unbounded we can find an ordinal $f(0) \in A$, $f(0) > \alpha$. Since B is unbounded we can now find an ordinal $f(1) \in B$ with $f(1) > f(0)$. In general we can always find $f(2n+2) \in A$ with $f(2n+2) > f(2n+1)$, and then $f(2n+3) \in B$ with $f(2n+3)$

$> f(2n+2)$. Let $\beta = \sup f[\omega]$. Clearly, $\beta = \sup_{n<\omega} f(2n) = \sup_{n<\omega} f(2n+1)$. But $f(2n) \in A$ for all n and $f(2n+1) \in B$ for all n. Thus $\beta \in A$ and $\beta \in B$. Hence $\beta \in A \cap B$ and we are done.

The reader might like to check that whenever $C \subseteq \kappa$ is closed and unbounded (κ an uncountable regular cardinal), then $C' = \{\alpha \in C \mid \alpha$ is a limit ordinal $\}$ is also closed and unbounded.

A subset E of a regular uncountable cardinal κ is said to be <u>stationary</u> if $E \cap C \neq \emptyset$ for all closed and unbounded sets $C \subseteq \kappa$. By our above result, every closed unbounded set will be stationary. The converse is easily disproved. (Remove one limit point from a closed unbounded set.) Indeed, we can find disjoint stationary subsets of κ. However, a stationary set is certainly unbounded (in κ), since the set $\{\beta \in \kappa \mid \beta > \alpha\}$ is closed and unbounded for any $\alpha < \kappa$. For example, E and F are disjoint stationary subsets of ω_2, where

$$E = \{\alpha \in \omega_2 \mid cf(\alpha) = \omega\} \quad , \quad F = \{\alpha \in \omega_2 \mid cf(\alpha) = \omega_1\} .$$

We are now ready to introduce one of Jensen's combinatorial principles. The one we consider is both the simplest and the most widely used so far. (These two aspects may, of course, be related.) After Jensen, we denote this principle by $\diamondsuit$ (i.e. <u>diamond</u>).

$\diamondsuit$: Let κ be an uncountable regular cardinal. Then there is a sequence $\{S_\alpha\}_{\alpha<\kappa}$ such that each S_α is a subset of α and whenever $X \subseteq \kappa$ there is a stationary set $E \subseteq \kappa$ such that $X \cap \alpha = S_\alpha$ for all $\alpha \in E$.

(The symbol $\diamondsuit$ is sometimes used to denote the specific instance of this principle when $\kappa = \omega_1$, with $\diamondsuit_\kappa$ then being used to denote the version for any other κ.

Assuming V = L, Jensen proved $\diamondsuit$. So let us see what $\diamondsuit$ says. Well now we know that there are at least κ^+ many subsets of κ. $\diamondsuit$ says that there are

κ many sets $S_\alpha \subseteq \alpha$ such that on a stationary set S_α predicts the correct value of the initial part $X \cap \alpha$ of any set $X \subseteq \kappa$. Since any stationary set is unbounded, we see that in fact the S_α's give arbitrarily good approximations to X, and that with E the appropriate stationary set, $X = \bigcup_{\alpha \in E}(X \cap \alpha) = \bigcup_{\alpha \in E} S_\alpha$. Now, readers familiar with GCH will observe at once that if we assume GCH (a weaker assumption than V = L), we can define a sequence $\{S_\alpha\}_{\alpha<\kappa}$ such that $S_\alpha \subseteq \alpha$ for all $\alpha<\kappa$ and whenever $X \subseteq \kappa$, then for each $\alpha<\kappa$ there is a $\beta<\kappa$, $\beta \geqslant \alpha$, with $X \cap \alpha = S_\beta$. What $\Diamond$ says is that we can find such a sequence with the property that on a stationary set of α's the β here can be taken to be α itself. This turns out to be much stronger than the GCH result. Indeed, Jensen has proved that $\Diamond$ does not follow from GCH.

Suprisingly, the fact that every set X gets approximated on a <u>stationary</u> (i.e. very large) set of α's in the manner $X \cap \alpha = S_\alpha$ in $\Diamond$ is irrelevant. Working in the system ZFC, <u>Devlin</u> has shown that $\Diamond_\kappa$ is equivalent to the existence of a sequence $\{S_\alpha\}_{\alpha<\kappa}$ with $S_\alpha \subseteq \alpha$ such that whenever $X \subseteq \kappa$ there is <u>at least one</u> infinite ordinal α for which $X \cap \alpha = S_\alpha$. (In many applications of $\Diamond$ this is all that is required.) Thus the essence of $\Diamond$ is not that we can approximate any set arbitrarily well, but that we can approximate it at least once <u>in the manner</u> $X \cap \alpha = S_\alpha$. This does not mean, however, that we never make use of the stationary set in $\Diamond$. Indeed, in our first application of $\Diamond$ below we make essential use of the fact that the approximations are good on a stationary set.

2. <u>The Souslin Problem</u>

We solve the Souslin Problem by proving the following theorem:

2.1 <u>Theorem</u>

Assume V = L. Then there is a linearly ordered set $(Y,\leqslant)$ such that:

(i) $\leqslant$ is a dense linear ordering without end points;

(ii) Y satisfies the c.c.c. (i.e. there is no uncountable family of disjoint open intervals);

(iii) $(Y,\leq)$ is Dedekind complete (i.e. every bounded subset of Y has a lub and a glb) ;

(iv) $(Y,\leq)$ is not isomorphic to the real line, $\mathbb{R}$. □

We first observe that it is sufficient to prove the following result:

2.2 Lemma

Assume V = L. Then there is a linearly ordered set $(X,\leq)$ such that:

(i) $\leq$ is a dense linear ordering without end points ;

(ii) Every closed and nowhere dense subset of X is countable ;

(iii) X is not separable (i.e. X has no countable dense subset) . □

We are thus claiming that 2.2 implies 2.1. To see this, let $(X,\leq)$ be as in 2.2. Let $(Y,\leq)$ be the Dedekind completion of X. It is immediate that Y satisfies conditions (i) and (iii) of 2.1. Suppose it failed to satisfy 2.1(ii). It follows at once that X fails to satisfy the c.c.c. as well. So let $\{(a_i,b_i) \mid i \in I\}$ be an uncountable family of disjoint open intervals of X. By Zorn's lemma we may assume this family is maximal. Set $K = X - \bigcup_{i\in I}(a_i,b_i)$. Clearly, K is closed and nowhere dense in X. But $\{a_i \mid i \in I\} \subseteq K$ so K is uncountable. This violates 2.2(ii). Hence Y satisfies 2.1(i)-(iii). Suppose it failed to satisfy 2.1(iv). Thus $(Y,\leq) \cong \mathbb{R}$. Hence X is isomorphic to a subspace of $\mathbb{R}$. But this implies that X is separable, contrary to 2.2(iii).

We assume V = L from now on. We construct an ordered set $(X,\leq)$ as in 2.2. The construction is by induction. Clearly, it is the ordering on the points of X which is important, not what those points are. So we shall take for the set X the set ω_1 of all countable ordinals. We define first the ordering on ω , then its extension to $\omega+\omega$, then the extension of that ordering to $\omega+\omega+\omega$, etc. In the following discussion, α and β will always denote limit ordinals less than ω_1.

By induction on α we define an ordering $<^*_\alpha$ on α so that:

(i) $<^*_\alpha$ is a dense linear ordering without end points;

(ii) if $\alpha < \beta$, then $<^*_\alpha = <^*_\beta \cap (\alpha \times \alpha)$.

(In connection with (ii), recall that any binary relation is a set of ordered pairs, so $<^*_\alpha$ will always be a subset of the Cartesian product $\alpha \times \alpha$.)

To commence we take $<^*_\omega$ to be any ordering of ω isomorphic to $\mathbb{Q}$, the rational line. And if β is a limit of limit ordinals we shall set $<^*_\beta = \bigcup_{\alpha<\beta} <^*_\alpha$, which by (ii) will satisfy (i) and (ii). And when we are through we shall set $<^* = \bigcup_{\alpha < \omega_1} <^*_\alpha$ to obtain our desired ordering of ω_1 (i.e. X). It will be immediate that $<^*$ satisfies 2.2(i). So we must define $<^*_{\beta+\omega}$ from $<^*_\beta$ in each case to ensure (somehow) that both 2.2(ii) and 2.2(iii) will hold.

Ensuring that 2.2(iii) will hold is easy. It suffices that we arrange that for no $\beta < \omega_1$ is β ($= \{\nu | \nu < \beta\} \subseteq \omega_1$) dense in $(\omega_1, <^*)$. Why? Well, since $\mathrm{cf}(\omega_1) = \omega_1$, if $A \subseteq \omega_1$ is countable, then $A \subseteq \beta$ for some $\beta < \omega_1$, so if A were dense in $(\omega_1, <^*)$, so would β be dense in $(\omega_1, <^*)$. So how do we arrange that no β will be dense in $(\omega_1, <^*)$? Suppose we are about to define $<^*_{\beta+\omega}$ from $<^*_\beta$. Let D_β be a proper Dedekind cut in $(\beta, <^*_\beta)$ (i.e. a proper initial segment of $(\beta, <^*_\beta)$ with no lub). Since $(\beta, <^*_\beta)$ is a countable, densely linearly ordered set, it will be isomorphic to $\mathbb{Q}$, so there is no difficulty in finding such a D_β. We define $<^*_{\beta+\omega}$ by ordering the ordinals $\beta+n$ $(n \in \omega)$ isomorphically to $\mathbb{Q}$ and placing this entire copy of $\mathbb{Q}$ where the "hole" at the top of D_β is. Thus, for each $n \in \omega$ we shall have

$$D_\beta \quad <^*_{\beta+\omega} \quad \beta+n \quad <^*_{\beta+\omega} \quad \beta - D_\beta \quad .$$

Thus β will not be dense in $\beta+\omega$, and hence certainly not in ω_1.

We now show that by choosing the cuts D_β carefully now, we can also ensure that 2.2(ii) will hold. The idea is to apply the Baire category theorem to avoid certain closed and nowhere dense sets. We first make precise what is meant here.

Let K be a closed and nowhere dense subset of $\mathbb{Q}$. Let D be a proper Dedekind cut in $\mathbb{Q}$. We say D <u>avoids</u> K if there are rationals $a < b$ with $a \in D < b$ and $[a,b] \cap K = \emptyset$.

2.3 <u>Lemma</u>

Let K_n, $n = 0,1,2,\ldots$ be closed and nowhere dense subsets of $\mathbb{Q}$. Then there is a proper Dedekind cut which avoids each K_n.

Proof: This is really just a special case of the Baire category theorem. The closures of the sets K_n in $\mathbb{R}$ are closed and nowhere dense subsets of $\mathbb{R}$, so there is an irrational x which is not in the closure of any of the sets K_n. Take D to be the cut defined by x. □

Since $(\beta, <^*_\beta) \cong \mathbb{Q}$, 2.3 will apply to this ordering. What we must specify then is which closed and nowhere dense subsets of β we are going to avoid. This is where we make use of V = L. In fact, what we use is the combinatorial principle $\Diamond$.

By $\Diamond$, let $\{S_\alpha\}_{\alpha<\omega_1}$ be a sequence such that $S_\alpha \subseteq \alpha$ and whenever $A \subseteq \omega_1$ the set $\{\alpha \in \omega_1 \mid A \cap \alpha = S_\alpha\}$ is stationary in ω_1.

In order to define $<^*_{\beta+\omega}$ now, we use 2.3 to pick D_β to avoid all those sets S_α, $\alpha \leq \beta$, which happen to be closed and nowhere dense subsets of β.

The definition of the ordered set thus completed, it remains only to prove that it is as required by 2.2. In view of our earlier remarks we need only check 2.2(ii) now.

Let $K \subseteq \omega_1$ be closed and nowhere dense in ω_1 (under $<^*$). We must prove that K is countable.

The idea is to find a countable ordinal α for which $K = K \cap \alpha$. This is where we make use of $\Diamond$. By defining a suitable closed unbounded set $C \subseteq \omega_1$ and picking any $\alpha \in C$ for which $K \cap \alpha = S_\alpha$, we shall be able to find such an α.

Define functions f, g from ω_1 to ω_1 so that for each $\nu \in \omega_1 - K$,

$$f(\nu) <^* \nu <^* g(\nu) \quad \text{and} \quad (f(\nu), g(\nu))^* \cap K = \emptyset.$$

Define functions h, k from $\omega_1 \times \omega_1$ to ω_1 so that whenever $\nu <^* \tau$,

$$h(\nu,\tau),\ k(\nu,\tau) \in (\nu,\tau)^* ;$$

$$h(\nu,\tau) \notin K ;$$

$$k(\nu,\tau) \in K \text{ whenever } K \cap (\nu,\tau)^* \neq \emptyset.$$

Set $C = \{\alpha \in \omega_1 \mid \alpha$ is a limit ordinal and for all $\nu < \alpha$, $f(\nu) < \alpha$ and $g(\nu) < \alpha$, and for all $\nu, \tau < \alpha$, $h(\nu,\tau) < \alpha$ and $k(\nu,\tau) < \alpha\}$.

2.4 Lemma

(i) C is closed and unbounded in ω_1;

(ii) $\alpha \in C \rightarrow \alpha$ is a limit ordinal;

(iii) $\alpha \in C \rightarrow K \cap \alpha$ is closed and nowhere dense in α.

Proof: It is an instructive exercise to prove part (i). (Set theorists will see this immediately. For the non-expert, notice that closure is trivial, and to prove the unboundedness of C use the fact that the limit of any monotone increasing sequence from ω_1 is a countable limit ordinal.) And part (ii) is included in the definition of C. For part (iii), if $\alpha \in C$, then since α is closed under f and g, $K \cap \alpha$ will be closed in α (under $<^*_\alpha$), and since α is closed under h, $K \cap \alpha$ will be nowhere dense in α. □

By $\diamondsuit$, fix now some ordinal $\alpha \in C$ such that $K \cap \alpha = S_\alpha$.

2.5 Lemma

Let $\beta \geq \alpha$. Then (i) $K \cap \beta = K \cap \alpha$, and (ii) if $\gamma \in \beta - K$ there are $\nu, \tau \in \alpha$ with $\nu <^* \gamma <^* \tau$ and $(\nu,\tau)^* \cap K = \emptyset$.

Proof: By induction on β. For $\beta = \alpha$ we use the closure of α under f and g. And the induction step at limit stages is trivial. So assume now the result holds for β. We prove it for $\beta+\omega$. By the induction hypothesis, $S_\alpha = K\cap\beta$ is closed and nowhere dense in β, so by construction, D_β avoids S_α. Thus there are $\gamma <^* \gamma'$ in β such that $[\gamma,\gamma']^* \cap K \cap \beta = \emptyset$ and all the ordinals $\beta+n$ $(n\in\omega)$ lie in $(\gamma,\gamma')^*$. Let $\nu,\tau,\nu',\tau' < \alpha$ arise from applying part (ii) of the induction hypothesis (i.e. part (ii) of the lemma) to γ,γ', respectively. Then $(\nu,\tau')^* \cap K \cap \alpha = \emptyset$, so as α is closed under the function k, $(\nu,\tau')^* \cap K = \emptyset$. But since all the ordinals $\beta+n$ $(n\in\omega)$ lie in $(\nu,\tau')^*$, (i) and (ii) follow immediately at $\beta+\omega$. The proof is complete. □

By (i) of 2.5 now, we have $K = \bigcup_\beta (K\cap\beta) = K\cap\alpha$. Hence $K\subseteq\alpha$. In particular, K is countable. The proof of 2.2 is now complete.

3. The Whitehead Problem

Assuming V = L, we prove that if G is an abelian group for which

$$\mathrm{Ext}(G,\mathbb{Z}) = 0,$$

then G is free. Just as in I.2, we use the word "group" to mean "abelian group". We commence with some preliminaries from group theory and homological algebra.

Let $G \xrightarrow{\varphi} B$ be a group epimorphism. A split of φ is a homomorphism $B \xrightarrow{\psi} G$ such that $\varphi\circ\psi = 1_B$. Any such ψ must clearly be a monomorphism.

Let G be an extension of A by B. We say G is a split extension just in case the canonical projection $G \xrightarrow{\varphi} B$ has a split. In this case, if $B \xrightarrow{\psi} G$ is a split of φ, it is easily seen that $G = A \oplus \mathrm{Im}(\psi)$, so G "splits" into the two components A and (up to isomorphism) B.

It follows from the above that $\mathrm{Ext}(G,\mathbb{Z}) = 0$ iff every extension of $\mathbb{Z}$ by G is a split extension. We may also characterise the property $\mathrm{Ext}(G,\mathbb{Z}) = 0$ in terms of short exact sequences.

A <u>sequence</u> (of morphisms)

$$A_1 \xrightarrow{\varphi_1} A_2 \xrightarrow{\varphi_2} \cdots \xrightarrow{\varphi_{n-1}} A_n \xrightarrow{\varphi_n} A_{n+1}$$

is <u>exact</u> if $\mathrm{Im}(\varphi_i) = \mathrm{Ker}(\varphi_{i+1})$ for all $i = 1,\ldots,n-1$.

For example, the sequence

$$0 \longrightarrow A \xrightarrow{\pi} G$$

is exact iff π is one-to-one. And the sequence

$$G \xrightarrow{\varphi} B \longrightarrow 0$$

is exact iff φ is onto.

A <u>short sequence</u> is one of the form

$$0 \longrightarrow A \xrightarrow{\pi} G \xrightarrow{\varphi} B \longrightarrow 0.$$

It will be exact iff G is an extension of $\mathrm{Im}(\pi)$ by B , in which case φ will be the canonical projection.

A short exact sequence

$$0 \longrightarrow A \xrightarrow{\alpha} B \xrightarrow{\beta} C \longrightarrow 0$$

<u>splits</u> if there is a $B \xrightarrow{\gamma} A$ with $\gamma \circ \alpha = 1_A$. This will be the case iff there is a $C \xrightarrow{\delta} B$ with $\beta \circ \delta = 1_C$. (Proving this is an easy exercise on exactness.) And of course, this at once implies that we have just found another way of saying that B is a split extension of A by C.

Thus, $\mathrm{Ext}(G,\mathbb{Z}) = 0$ iff every short exact sequence

$$0 \longrightarrow \mathbb{Z} \longrightarrow H \longrightarrow G \longrightarrow 0$$

splits.

So far we have succeeded in expressing some remarkably trivial facts in a very impressive looking language. But we at once achieve the payoff, since we may now apply the following fundamental result from homological algebra.

3.1 Lemma

Let

$$0 \longrightarrow A \xrightarrow{\lambda} B \xrightarrow{\mu} C \longrightarrow 0$$

be exact. Let G be any group. Then there is an exact sequence

$$0 \longrightarrow \mathrm{Hom}(C,G) \xrightarrow{\mu^*} \mathrm{Hom}(B,G) \xrightarrow{\lambda^*} \mathrm{Hom}(A,G) \longrightarrow \mathrm{Ext}(C,G) \longrightarrow \mathrm{Ext}(B,G) \longrightarrow$$

$$\longrightarrow \mathrm{Ext}(A,G) \longrightarrow 0 ,$$

where, in particular, $\lambda^*(\varphi) = \varphi \circ \lambda$ and $\mu^*(\psi) = \psi \circ \mu$. □

A proof of 3.1 may be found in almost any book on homological algebra. It is not difficult once one has a characterisation of Ext(H,K) in terms of equivalence classes of short exact sequences. But the proof is very tedious, and at any rate outside the scope of this book, so we leave it to the reader to investigate the matter himself.

We are now in a position to start our proof. We shall call a group G a W-group if $\mathrm{Ext}(G,\underset{\sim}{Z}) = 0$. By I.2.2, if G is free, then G is a W-group. It is our task to prove the converse. We first of all need three facts about W-groups.

3.2 Lemma

Let G be a W-group. If $H \lhd G$, then H is a W-group.

Proof: Let

$$0 \longrightarrow H \xrightarrow{1_H} G \xrightarrow{\varphi} G/H \longrightarrow 0$$

be exact. By 3.1 there is an exact sequence

$$\mathrm{Ext}(G,\underset{\sim}{Z}) \longrightarrow \mathrm{Ext}(H,\underset{\sim}{Z}) \longrightarrow 0.$$

Since G is a W-group,

$$\mathrm{Ext}(G,\underset{\sim}{Z}) = 0.$$

Hence, by exactness,

$$\mathrm{Ext}(H,\underset{\sim}{Z}) = 0,$$

and we are done. □

3.3 Lemma

If G is a W-group, then G is torsion free.

Proof: Suppose G has torsion. Then for some $g \in G$, $\langle g \rangle$ is finite, say of order n. Then $\langle g \rangle \cong \mathbb{Z}/n\mathbb{Z}$. But by 3.2, $\langle g \rangle$ will be a W-group. Hence $\mathbb{Z}/n\mathbb{Z}$ is a W-group. Let $\mathbb{Z} \xrightarrow{\pi} \mathbb{Z}/n\mathbb{Z}$ be the canonical projection. Clearly, $\mathrm{Ker}(\pi) \cong \mathbb{Z}$. Thus there is an exact sequence

$$0 \longrightarrow \mathbb{Z} \longrightarrow \mathbb{Z} \xrightarrow{\pi} \mathbb{Z}/n\mathbb{Z} \longrightarrow 0.$$

Since $\mathrm{Ext}(\mathbb{Z}/n\mathbb{Z},\mathbb{Z}) = 0$, this sequence must split. Hence there is an embedding

$$\mathbb{Z}/n\mathbb{Z} \longrightarrow \mathbb{Z}.$$

But $\mathbb{Z}/n\mathbb{Z}$ has torsion and $\mathbb{Z}$ is torsion free, so this is impossible. □

3.4 Lemma

Let $H \lhd G$. Suppose that G is a W-group but that G/H is not a W-group. Then there is a $H \xrightarrow{\varphi} \mathbb{Z}$ which does not extend to a $G \xrightarrow{\psi} \mathbb{Z}$.

Proof: Let

$$0 \longrightarrow H \xrightarrow{1_H} G \longrightarrow G/H \longrightarrow 0$$

be exact. By 3.1 there is an exact sequence

$$\mathrm{Hom}(G,\mathbb{Z}) \xrightarrow{1_H^*} \mathrm{Hom}(H,\mathbb{Z}) \xrightarrow{\rho} \mathrm{Ext}(G/H,\mathbb{Z}) \xrightarrow{\tau} \mathrm{Ext}(G,\mathbb{Z}),$$

where $1_H^*(\varphi) = \varphi \circ 1_H = \varphi$ restricted to H, for all $\varphi \in \mathrm{Hom}(G,\mathbb{Z})$.

Now, by hypothesis on G/H, $\mathrm{Ext}(G/H,\mathbb{Z}) \neq 0$. And by hypothesis on G, $\mathrm{Ext}(G,\mathbb{Z}) = 0$. Hence, $\mathrm{Im}(\rho) = \mathrm{Ker}(\tau) = \mathrm{Ext}(G/H,\mathbb{Z}) \neq 0$. Hence $\mathrm{Ker}(\rho) \neq \mathrm{Hom}(H,\mathbb{Z})$. Hence $\mathrm{Im}(1_H^*) \neq \mathrm{Hom}(H,\mathbb{Z})$. This means that there is a $H \xrightarrow{\varphi} \mathbb{Z}$ with $\varphi \neq 1_H^*(\psi)$ for any $G \xrightarrow{\psi} \mathbb{Z}$. In other words, φ is not the restriction to H of any $G \xrightarrow{\psi} Z$. Thus φ is as required. □

Suppose now that we are given a W-group, G, and we wish to show that G is free. How might we proceed? One way is to try to construct, explicitly, a basis for G. A useful tool then would be the following result, a direct consequence of the proof of I.2.1.

3.5 Lemma

Let $H \lhd G$. If G/H and H are both free, then G is free and any basis of H can be extended to a basis of G. □

We might try to use 3.5 as follows. Suppose we can represent G as the union of an increasing, continuous chain $\{G_\nu\}_{\nu<\tau}$ of subgroups, for some limit ordinal τ. (Increasing here means that $\nu<\nu'$ implies $G_\nu \lhd G_{\nu'}$. Continuous means that for any limit ordinal $\sigma<\tau$, $G_\sigma = \bigcup_{\nu<\sigma} G_\nu$.) Then we could try first to find a basis for G_0, then extend it to a basis for G_1, then extend that to a basis for G_2, and so on. 3.6 provides us with a sufficient condition for this process to work:

3.6 Lemma

Let $G = \bigcup_{\nu<\tau} G_\nu$, where τ is a limit ordinal and $\{G_\nu\}_{\nu<\tau}$ is an increasing, continuous chain of subgroups of G. Suppose G_0 is free and that $G_{\nu+1}/G_\nu$ is free for all $\nu<\tau$. Then G is free, and so is each group G/G_ν.

Proof: By induction, we construct an increasing (under inclusion) sequence $\{X_\nu\}_{\nu<\tau}$ of sets such that X_ν is a basis for G_ν. If $\sigma<\tau$ is a limit ordinal we shall have $X_\sigma = \bigcup_{\nu<\sigma} X_\nu$. At successor stages we obtain $X_{\nu+1}$ from X_ν by 3.5. Clearly, if we set $X = \bigcup_{\nu<\tau} X_\nu$, X will be a basis for G. Moreover, for any $\nu<\tau$, the set

$$\{ G_\nu + x \mid x \in X - X_\nu \}$$

is a basis for G/G_ν. □

If G is a torsion free group, we shall say that a subgroup H of G is pure if G/H is torsion free. Using 3.6, we obtain a useful condition for a group to be free for the case that the group concerned is countable.

3.7 Lemma (Pontryagin's Criterion)

Let G be a countable, torsion free group. If every finitely generated subgroup of G is contained in a finitely generated pure subgroup of G, then G is free.

Proof: Let $G = \{g_n \mid n \in \omega\}$. Define a sequence $\{G_n\}$ of subgroups of G as follows. Let $G_0 = 0$. Suppose G_n has been defined and is finitely generated. Then $\langle G_n, g_n \rangle$ is finitely generated, so by assumption there is a finitely generated pure subgroup of G containing this group. Let G_{n+1} be such a group.

Clearly, $G = \bigcup_{n<\omega} G_n$. For each n, G_n is pure in G, so G/G_n is torsion free. Hence G_{n+1}/G_n is torsion free for each n. But G_{n+1} is finitely generated, each n, whence so is G_{n+1}/G_n. But it is a standard result of group theory that any finitely generated torsion free group is free. So, as G_0 is free, 3.6 tells us that G is free.□

We shall prove our main result that (assuming V = L) every W-group is free by induction on the order of G. For the case of countable groups we shall use 3.7. For the general case we use a generalisation of 3.7 due to S. U. Chase.

Suppose now that we are given a group B and we wish to construct, explicitly, an extension C of $\mathbb{Z}$ by B. As domain for our extension it is natural to take the Cartesian product $B \times \mathbb{Z}$. We must therefore define a group operation on this set. Since we want C to be an extension of $\mathbb{Z}$ we shall require σ to embed $\mathbb{Z}$ in C, where $\sigma : \mathbb{Z} \to B \times \mathbb{Z}$ is defined by $\sigma(n) = (0,n)$. And since C should extend $\mathbb{Z}$ by B, we shall demand that $\pi : B \times \mathbb{Z} \to B$ is a group homomorphism, where $\pi((b,n)) = b$. This will imply that $\mathrm{Ker}(\pi) = \mathrm{Im}(\sigma)$, so the sequence

$$0 \longrightarrow \mathbb{Z} \xrightarrow{\sigma} C \xrightarrow{\pi} B \longrightarrow 0$$

will indeed be exact.

We say C is a $(B,\mathbb{Z})$-group iff:-

(i) C has domain $B \times \mathbb{Z}$;

(ii) $\sigma : \mathbb{Z} \to C$ is a group morphism, where $\sigma(n) = (0,n)$;

(iii) $\pi : C \to B$ is a group morphism, where $\pi((b,n)) = b$.

In the context of $(B,\mathbb{Z})$-groups, the symbols σ, π will always have the above meanings.

One example of a $(B,\underset{\sim}{Z})$-group is the external direct sum $B \dot{\oplus} \underset{\sim}{Z}$, but as we shall see this is not the only example.

3.8 Lemma

Let B_1 be a W-group, $B_0 \lhd B_1$. Let C_0 be a torsion free $(B_0,\underset{\sim}{Z})$-group. Then there is a torsion free $(B_1,\underset{\sim}{Z})$-group C_1 such that $C_0 \lhd C_1$.

Proof: Since $B_0 \lhd B_1$, B_0 is a W-group. Let $B_0 \xrightarrow{\rho_0} C_0$ split $C_0 \xrightarrow{\pi_0} B_0$. Define $\tau : B_0 \dot{\oplus} \underset{\sim}{Z} \longrightarrow C_0$ by $\tau((b,n)) = \rho_0(b) + (0,n)$. It is easily checked that τ is a group isomorphism. Moreover, $\pi_0 \circ \tau((b,n)) = b$ and $\tau((0,n)) = (0,n)$. Hence we may assume, without loss of generality, that τ is the identity here, and that $C_0 = B_0 \dot{\oplus} \underset{\sim}{Z}$. Then $C_1 = B_1 \dot{\oplus} \underset{\sim}{Z}$ is as required. □

3.9 Lemma

Let B_1 be a W-group, $B_0 \lhd B_1$. Suppose that B_1/B_0 is not a W-group. Let C_0 be a torsion free $(B_0,\underset{\sim}{Z})$-group, and suppose that $B_0 \xrightarrow{\rho_0} C_0$ splits $C_0 \xrightarrow{\pi_0} B_0$. Then there is a torsion free $(B_1,\underset{\sim}{Z})$-group C_1 such that $C_0 \lhd C_1$ and ρ_0 does not extend to a split of $C_1 \xrightarrow{\pi_1} B_1$.

Proof: As in 3.8 we may assume that $\tau = 1_{B_0 \dot{\oplus} \underset{\sim}{Z}}$ and $C_0 = B_0 \dot{\oplus} \underset{\sim}{Z}$, where we define τ by $\tau((b,n)) = \rho_0(b) + (0,n)$. In particular, since $\rho_0(b) = \tau((b,0))$, we shall now have $\rho_0(b) = (b,0)$.

By 3.4, let $B_0 \xrightarrow{\psi} \underset{\sim}{Z}$ have no extension to B_1. Set $\tilde{C}_1 = B_1 \dot{\oplus} \underset{\sim}{Z}$, and define a function $\gamma : C_0 \longrightarrow \tilde{C}_1$ by

$$\gamma((b,n)) = (b,n+\psi(b)).$$

Clearly, γ is a group morphism. We prove that:

(*) There is no $B_1 \xrightarrow{\tilde{\rho}} \tilde{C}_1$ such that the restriction of $\tilde{\rho}$ to B_0 is $\gamma \circ \rho_0$.

In order to prove (*), we assume the contrary, i.e.that there were such a $\tilde{\rho}$.

Define $\theta : \tilde{C}_1 \longrightarrow \underset{\sim}{Z}$ by

$$\theta((b,n)) = n ,$$

and define $\varphi : B_1 \longrightarrow \underset{\sim}{Z}$ by

$$\varphi = \theta \circ \tilde{\rho} .$$

Then θ and φ are group morphisms and for any $b \in B_0$,

$$\varphi(b) = \theta \circ \tilde{\rho}(b) = \theta \circ \gamma \circ \rho_0(b) = \theta \circ \gamma((b,0)) = \theta((b,\psi(b))) = \psi(b) .$$

Hence φ extends ψ on B_1 , which is impossible. This proves (*).

Now define a function $f : \tilde{C}_1 \longrightarrow B_1 \times \underset{\sim}{Z}$ by

$$f((b,n)) = \begin{cases} (b,n) & , \text{ if } b \notin B_0 \\ (b,n-\psi(b)) & , \text{ if } b \in B_0 . \end{cases}$$

Clearly, f is a bijection. Let C_1 be the group on $B_1 \times \underset{\sim}{Z}$ induced from $\tilde{C}_1$ by f. Since $f \circ \gamma = 1_{B_0 \times \underset{\sim}{Z}}$, $C_0 \lhd C_1$. Amd clearly, C_1 is a torsion free $(B_1,\underset{\sim}{Z})$-group.

Suppose $B_1 \xrightarrow{\rho_1} C_1$ were to split $C_1 \xrightarrow{\pi_1} B_1$, where the restriction of ρ_1 to B_0 is ρ_0 . Set $\tilde{\rho} = f^{-1} \circ \rho_1$. Thus $B_1 \xrightarrow{\tilde{\rho}} \tilde{C}_1$ and for $b \in B_0$,

$$\tilde{\rho}(b) = f^{-1} \circ \rho_1(b) = f^{-1} \circ \rho_0(b) = f^{-1}((b,0)) = (b,\psi(b)) = \gamma((b,0)) = \gamma \circ \rho_0(b).$$

This contradicts (*) . Hence C_1 is as required. □

The following result was proved by <u>K. Stein</u> in 1951. There are much more direct proofs of this result, but the one given here has the advantage that it will generalise to the uncountable case. The only difference between the two cases is that, whereas the countable case is provable in ZFC, we need V = L in order to carry through the proof in the uncountable case.

3.10 Theorem

Every countable W-group is free.

Proof: Let G be a countable W-group. By 3.3, G is torsion free. So by 3.7, in order to prove that G is free it suffices to prove that G satisfies Pontryagin's criterion. We assume it does not and work for a contadiction.

Let B_0 be a finitely generated subgroup of G which is not contained in any finitely generated pure subgroup of G. Let

$$B = \{g \in G \mid (\exists k \in \omega)(k > 0 \ \&\ k.g \in B_0)\} .$$

B is a (countable) subgroup of G. And by its very definition, B is pure in G. Since $B_0 \subseteq B$, B cannot, therefore, be finitely generated. Hence we can represent B as a union of a strictly increasing chain $\{B_n\}$ of finitely generated subgroups, commencing with B_0. Notice that as $B_0 \neq B$, the definition of B implies that B/B_0 is a torsion group. Let S be a finite set of generators for B_0.

By induction on n we construct a strictly increasing sequence $\{C_n\}$ of groups such that each C_n is a torsion free $(B_n, \underset{\sim}{Z})$-group and $C = \bigcup_{n<\omega} C_n$ is a torsion free $(B, \underset{\sim}{Z})$-group. The idea is to construct the C_n's so that $C \xrightarrow{\pi} B$ does not split, thereby contadicting the fact that by 3.2, $B \lhd G$ is a W-group. We need to know the following fact.

(*) If C is a torsion free group and $B \xrightarrow{\rho} C$, then ρ is uniquely determined by the values it takes on S.

Proof of (*): Let $b \in B$. Pick $n > 0$ with $n.b \in B_0$. Since S generates B_0, $\rho(n.b)$ is determined by the values of ρ on S. Thus $n.\rho(b)$ is determined by the values of ρ on S. But C is torsion free, so if $n.c = n.\rho(b)$, then $c = \rho(b)$. Hence $\rho(b)$ is uniquely determined by the values ρ takes on S. QED(*).

Fix now some enumeration $\{g_n\}$ of all functions $g:S \longrightarrow S \times \mathbb{Z}$ with $\pi \cdot g = 1_S$.

Set $C_0 = B_0 \dot{\oplus} \mathbb{Z}$. Suppose now that C_n has been defined. If g_n extends to a split ρ of $C_n \xrightarrow{\pi_n} B_n$, let C_{n+1} be an extension of C_n such that ρ does not extend to a split of $C_{n+1} \xrightarrow{\pi_{n+1}} B_{n+1}$. (Since B_{n+1}/B_n is a torsion group, it is not a W-group, so by 3.9 such a C_{n+1} can be found. Notice also that by (*), if there is such a ρ, it will be unique.) Otherwise use 3.8 to let C_{n+1} be any $(B_{n+1}, \mathbb{Z})$-group extending C_n. That completes the construction.

We finish by showing that $C \xrightarrow{\pi} B$ does not split. Suppose, on the contrary, that $B \xrightarrow{\rho} C$ were to split π. For some n, g_n must be the restriction of ρ to S. Let ρ_n be the restriction of ρ to B_n. Then ρ_n splits $C_n \xrightarrow{\pi_n} B_n$. But ρ_n is the unique extension of g_n to a split of π_n. Hence, by construction of C_{n+1}, ρ_n does not extend to a split of $C_{n+1} \xrightarrow{\pi_{n+1}} B_{n+1}$. But this is absurd, since ρ_{n+1} splits π_{n+1}. The proof is complete. □

Assuming V = L, we shall prove by induction on the infinite cardinal κ that if G is a W-group of order κ, then G is free. The above result establishes the initial case $\kappa = \aleph_0$. We consider next the case $\kappa = \aleph_1$. The argument we use will work equally well for the induction step at any regular cardinal κ, so we shall not bother to look at these cases. This will leave the case where κ is a singular cardinal (and hence necessarily a limit cardinal), we have proved the result for all cardinals less than κ, and we wish to prove the result at κ. It turns out that this has nothing to do with V = L. Rather, GCH (which holds because we are assuming V = L for our proof) suffices, together with the fact that any set of cardinality κ will be a union of fewer than κ sets each of which has cardinality less than κ. Since this case thus has nothing to do with V = L, we shall omit it. A proof appears in <u>Shelah</u>'s paper [Sh].

Let us call a group G κ-free if every subgroup of G of order less than κ is free. By 3.10 and 3.2, we have at once:

3.11 Lemma

Every W-group is $\aleph_1$-free. □

Now, a group G will be torsion free iff every finitely generated subgroup is free. Viewed in this light, we see that the property of a group being κ-free is a natural generalisation of the property of being torsion free. In particular, if G is κ-free, it is torsion free. Accordingly, we define a subgroup H of a κ-free group G to be κ-pure if G/H is κ-free. Generalising the Pontryagin criterion to the $\aleph_1$ case we now have:

Chase's Condition: G is an $\aleph_1$-free group and every countable subgroup of G is contained in a countable $\aleph_1$-pure subgroup of G.

The following result is trivial.

3.12 Lemma

Let G be a group of order $\aleph_1$. Then G satisfies Chase's condition iff G is the union of an increasing, continuous sequence $\{A_\nu\}_{\nu<\omega_1}$ of countable free groups with $A_0 = 0$ and, for all $\nu < \omega_1$, $A_{\nu+1}$ is $\aleph_1$-pure in G. □

Now, unlike in the case of the Pontryagin criterion, we cannot prove that any group which satisfies Chase's condition is free. However, we have a partial result in this direction:

3.13 Lemma

Let G be a group of order $\aleph_1$ which satisfies Chase's condition, and let $\{A_\nu\}_{\nu<\omega_1}$ be as in 3.12. Then G is free iff there is a closed and unbounded set $C \subseteq \omega_1$ such that A_ν is $\aleph_1$-pure in G for all $\nu \in C$.

Proof: ($\leftarrow$) Let $\{\alpha_\nu\}_{\nu<\omega_1}$ be the monotone enumeration of such a set C. Set

$$\tilde{A}_\nu = A_{\alpha_\nu}, \quad \text{each } \nu .$$

For each ν, $\tilde{A}_\nu$ is $\aleph_1$-pure in G, so $G/\tilde{A}_\nu$ is $\aleph_1$-free, so $\tilde{A}_{\nu+1}/\tilde{A}_\nu$ is free. So by 3.6 (applied to the sequence $\{\tilde{A}_\nu\}_{\nu<\omega_1}$) we see that G is free.

($\rightarrow$) Let X be a basis for G. By an easy induction we can define a closed and unbounded set $C \subseteq \omega_1$ and an increasing sequence $\{X_\nu\}_{\nu \in C}$ of subsets of X such that for each $\nu \in C$, X_ν is a basis of A_ν. Then, for each $\nu \in C$, the set

$$\{ A_\nu + x \mid x \in X - X_\nu \}$$

is a basis for G/A_ν. Hence G/A_ν is free for all $\nu \in C$, and hence is certainly $\aleph_1$-free. Thus each A_ν, $\nu \in C$, is $\aleph_1$-pure. □

In order to complete our proof that every W-group of order $\aleph_1$ is free (assuming V = L), we need a generalisation of the combinatorial principle $\diamondsuit$.

By a slight generalisation of the argument used to establish $\diamondsuit$ from the assumption V = L, one may prove the following (apparently) stronger result.

Let κ be an uncountable regular cardinal. Let $E \subseteq \kappa$ be stationary. Then there is a sequence $\{S_\alpha\}_{\alpha \in E}$ such that $S_\alpha \subseteq \alpha$ for each $\alpha \in E$, and whenever $X \subseteq \kappa$ there is a stationary set $E' \subseteq E$ such that $X \cap \alpha = S_\alpha$ for all $\alpha \in E'$.

This result is also due to Jensen. Shelah has proved that it does not follow from $\diamondsuit$ itself, though it clearly implies $\diamondsuit$.

Either by an entirely analogous argument, or else as a corollary of the above result, we may obtain the following consequence of V = L, tailor made for our present needs. (We give only the $\aleph_1$ case, since this is all we need now.)

3.14 Lemma

Assume V = L. Let B be the union of a strictly increasing, continuous sequence $\{B_\nu\}_{\nu<\omega_1}$ of countable sets. Let $E \subseteq \omega_1$ be stationary. Then there is a sequence $\{g_\nu\}_{\nu\in E}$ such that $g_\nu : B_\nu \longrightarrow B_\nu \times \underset{\sim}{Z}$ for all $\nu \in E$, and whenever $h: B \longrightarrow B \times \underset{\sim}{Z}$ is such that $h[B_\nu] \subseteq B_\nu \times \underset{\sim}{Z}$ for all $\nu \in E$, there is a $\nu \in E$ such that the restriction of h to B_ν is g_ν. □

3.15 Lemma

Assume V = L. Let B be a W-group of order $\aleph_1$. Let $\{B_\nu\}_{\nu<\omega_1}$ be a strictly increasing, continuous sequence of countable (free) subgroups of B with union B. Then there is a closed unbounded set $C \subseteq \omega_1$ such that $B_{\nu+1}/B_\nu$ is free for all $\nu \in C$.

Proof: Suppose not. Then the set

$$E = \{\nu \in \omega_1 \mid B_{\nu+1}/B_\nu \text{ is not free}\}$$

is stationary in ω_1. Let $\{g_\nu\}_{\nu\in E}$ be as in 3.14.

We define, by induction, a strictly increasing, continuous sequence $\{C_\nu\}_{\nu<\omega_1}$ of torsion free groups such that each C_ν is a $(B_\nu, \underset{\sim}{Z})$-group, $C = \bigcup_{\nu<\omega_1} C_\nu$ is a torsion free $(B, \underset{\sim}{Z})$-group, and $C \xrightarrow{\pi} B$ does not split, thereby contradicting the fact that B is a W-group.

Set $C_0 = B_0 \dot{\oplus} \underset{\sim}{Z}$. Suppose now that we have defined C_τ for all $\tau < \nu$. If ν is a limit ordinal we set $C_\nu = \bigcup_{\tau<\nu} C_\tau$. If $\nu = \tau + 1$, there are two cases to consider.

Suppose first that $\tau \in E$ and $B_\tau \xrightarrow{g_\tau} B_\tau \times \underset{\sim}{Z}$ splits $C_\tau \xrightarrow{\pi_\tau} B_\tau$. Then we let C_ν be an extension of C_τ such that g_τ does not extend to a split of $C_\nu \xrightarrow{\pi_\nu} B_\nu$. (Since $\tau \in E$, B_ν/B_τ is not free, so by 3.10 it is not a W-group. And as $B_\nu \lhd B$, B_ν is a W-group. So 3.9 guarantees the existence of such a C_ν.)

Otherwise we use 3.8 to let C_ν be any extension of C_τ.

Clearly, $C = \bigcup_{\nu<\omega_1} C_\nu$ is a torsion free $(B,\mathbb{Z})$-group. Suppose $B \xrightarrow{\rho} C$ were to split $C \xrightarrow{\pi} B$. Then we could find a $\tau \in E$ with g_τ equal to the restriction of ρ to B_τ. (Since ρ splits π, $\pi \circ \rho = 1_B$, so $\rho[B_\nu] \subseteq B_\nu \times \mathbb{Z}$ for all $\nu < \omega_1$.) Thus $C_{\tau+1}$ was defined so that no split of $C_{\tau+1} \xrightarrow{\pi_{\tau+1}} B_{\tau+1}$ agrees with ρ on B_τ. But this is absurd, since the restriction of ρ to $B_{\tau+1}$ splits $\pi_{\tau+1}$. □

3.16 Lemma

Assume V = L. If G is a W-group of order $\aleph_1$, then G satisfies Chase's condition.

Proof: Suppose not. Now, G is $\aleph_1$-free by 3.11, so it must be the case then that there is a countable subgroup B_0 of G such that whenever $C \lhd G$ is countable with $B_0 \subseteq C$, then C is not $\aleph_1$-pure. Pick such a B_0. Then, whenever $C \lhd G$ is countable and $B_0 \subseteq C$, we can find a countable $C' \lhd G$ with $C \lhd C'$ such that C'/C is not free. So by induction we can define a strictly increasing, continuous sequence $\{B_\nu\}_{\nu<\omega_1}$ of countable subgroups of G, commencing with B_0, such that $B_{\nu+1}/B_\nu$ is not free for each $\nu < \omega_1$. Set $B = \bigcup_{\nu<\omega_1} B_\nu$. Since $B \lhd G$, B is a W-group. And clearly, B has order $\aleph_1$. Thus by 3.15 there is a closed unbounded set $C \subseteq \omega_1$ such that $B_{\nu+1}/B_\nu$ is free for all $\nu \in C$. But this is absurd since for no ν is $B_{\nu+1}/B_\nu$ free. □

At last we may prove:

3.17 Theorem

Assume V = L. If G is a W-group of order $\aleph_1$, then G is free.

Proof: By 3.16 and 3.12 we may represent G as the union of a strictly increasing, continuous sequence $\{A_\nu\}_{\nu<\omega_1}$ of countable free groups such that $A_0 = 0$ and $A_{\nu+1}$ is $\aleph_1$-pure in G for each ν.

(*) For all $\nu < \omega_1$, A_ν is $\aleph_1$-pure in G iff $A_{\nu+1}/A_\nu$ is free.

Proof of (*): If A_ν is $\aleph_1$-pure in G, then G/A_ν is $\aleph_1$-free, so $A_{\nu+1}/A_\nu$ is free.

Conversely, Suppose $A_{\nu+1}/A_\nu$ is free. Let $\tau > \nu$. By the Third Isomorphism Theorem of elementary group theory,

$$(A_\tau/A_\nu)/(A_{\nu+1}/A_\nu) \cong A_\tau/A_{\nu+1} .$$

Now, $A_{\nu+1}$ is $\aleph_1$-pure in G, so $A_\tau/A_{\nu+1}$ is free. Hence, as $A_{\nu+1}/A_\nu$ is also free, 3.5 tells us that A_τ/A_ν is free. But $\tau > \nu$ is arbitrary here. Hence G/A_ν is $\aleph_1$-free. Hence A_ν is $\aleph_1$-pure in G . QED(*).

By 3.15 now, there is a closed and unbounded set $C \subseteq \omega_1$ such that $A_{\nu+1}/A_\nu$ is free for all $\nu \in C$. By (*), A_ν is $\aleph_1$-pure for all $\nu \in C$. Hence by 3.13, G is free. □

4. Collectionwise Hausdorff Spaces

It will be shown that if we assume V = L, every first countable T_4 space is collectionwise Hausdorff. The proof is by induction on the cardinality of the discrete set involved.

Let $(X,\mathfrak{T})$ be a topological space, κ an infinite cardinal. We say X is κ collectionwise Hausdorff (κ-CWH) if every discrete subset of X of cardinality κ has a separation.

The proof that, assuming V = L, every first countable T_4 space is collectionwise Hausdorff proceeds by establishing κ-CWH for all κ by induction on κ. For $\kappa = \aleph_0$ the result is provable in ZFC alone. The induction step at $\aleph_1$ is typical of the induction step at any uncountable regular cardinal. And the induction step at singular cardinals requires only the GCH, and depends upon the singularity of κ. So, just as with the Whitehead Problem, we give only the cases $\kappa = \aleph_0$ and $\kappa = \aleph_1$.

4.1 Theorem

Let X be any T_3 space. Then X is $\aleph_0$-CWH.

Proof: Let $Y \subseteq X$ be discrete, $|Y| = \aleph_0$. Notice that any discrete set is closed, and that any subset of a discrete set is discrete, and hence closed.

Enumerate Y as $\{y_n\}$. For each n, set

$$Y_n = \{y_{n+1}, y_{n+2}, \ldots\} \quad .$$

Now, $y_0 \notin Y_0$ and Y_0 is a closed subset of X, so by regularity there are disjoint open sets U_0, V_0 such that $y_0 \in U_0$ and $Y_0 \subseteq V_0$. Proceeding inductively, for each n we can gind disjoint open sets U_{n+1}, V_{n+1} such that $y_{n+1} \in U_{n+1}$, $Y_{n+1} \subseteq V_{n+1}$, and $U_{n+1}, V_{n+1} \subseteq V_n$.

The sets U_n, $n = 0,1,2, \ldots$ clearly separate Y. □

We turn next to the case $\kappa = \aleph_1$. Here we need V = L. Now, Jensen has proved that if V = L there is a sequence $\{f_\alpha\}_{\alpha<\omega_1}$ such that $f_\alpha : \alpha \to \omega$ for each α, and whenever $f: \omega_1 \to \omega$ there is a stationary set $E \subseteq \omega_1$ such that $f \restriction \alpha = f_\alpha$ for all $\alpha \in E$. (The symbol $f \restriction A$ denotes the restriction of f to A for any set $A \subseteq \mathrm{dom}(f)$.) Indeed, this is equivalent to the principle $\diamondsuit_{\aleph_1}$. For our present problem we need a somewhat stronger result:

4.2 Lemma

Assume V = L. Let A_f be, for each $f: \omega_1 \to \omega$, a stationary subset of ω_1, such that whenever $f: \omega_1 \to \omega$ and $g: \omega_1 \to \omega$ and $f \restriction \alpha = g \restriction \alpha$, then $A_f \cap (\alpha+1) = A_g \cap (\alpha+1)$. Then there is a sequence $\{f_\alpha\}_{\alpha<\omega_1}$ such that $f_\alpha : \alpha \to \omega+1$ for all α, and whenever $f: \omega_1 \to \omega$, the set $\{\alpha \in \omega_1 \mid f \restriction \alpha = f_\alpha\}$ is a stationary subset of A_f.□

Using 4.2, we prove:

4.3 Theorem

Assume V = L. Let X be any first countable T_4 space. Then X is $\aleph_1$-CWH.

Proof: Assume the theorem is false. Let Y be a discrete subset of X of cardinality $\aleph_1$ having no separation. We may without loss of generality assume that X is an ordinal and that Y is the ordinal ω_1. For each $x \in X$, let $\{N_n(x)\}$ enumerate a countable base for the neighborhood system of x.

If $f: \lambda \longrightarrow \omega$, where $\lambda \leq \omega_1$, and if $\alpha \leq \lambda$, we set

$$W(f,\alpha) = \bigcup_{\beta<\alpha} N_{f(\beta)}(\beta).$$

If $H \subseteq \omega_1$, $f: \lambda \longrightarrow \omega$, $\lambda \leq \omega_1$, and $\alpha \leq \lambda$, we set

$$W(H,f,\alpha) = \bigcup_{\beta \in H \cap \alpha} N_{f(\beta)}(\beta).$$

For $f: \omega_1 \longrightarrow \omega$ now, set

$$A_f = \{\alpha \in \omega_1 \mid (\underline{\text{closure}}\ W(f,\alpha)) \cap \omega_1 \neq \alpha\}.$$

Notice that if $f: \omega_1 \longrightarrow \omega$ and $g: \omega_1 \longrightarrow \omega$, then

$$f \upharpoonright \alpha = g \upharpoonright \alpha \quad \text{implies} \quad A_f \cap (\alpha + 1) = A_g \cap (\alpha + 1).$$

CLAIM: For each $f: \omega_1 \longrightarrow \omega$, A_f is stationary.

Proof of claim: Suppose not. Let $C \subseteq \omega_1$ be a closed and unbounded set which is disjoint from A_f (some f). Thus,

$$\alpha \in C \quad \text{implies} \quad (\underline{\text{closure}}\ W(f,\alpha)) \cap \omega_1 = \alpha.$$

Let $\{\alpha_\nu\}_{\nu<\omega_1}$ be the monotone enumeration of C. By 4.1, X is $\aleph_0$-CWH, so for each $\nu < \omega_1$ we can find a family $\mathfrak{F}_\nu$ of pairwise disjoint open sets U_x, $x \in \alpha_{\nu+1} - \alpha_\nu$, such that $x \in U_x$. We may clearly assume that $U_x \subseteq N_{f(x)}(x)$, each $x \in \omega_1$. Moreover, since $(\underline{\text{closure}}\ W(f,\alpha_\nu)) \cap \omega_1 = \alpha_\nu$ for each $\nu < \omega_1$, we may assume also that for $x > \alpha_\nu$, $U_x \cap$ closure $W(f,\alpha_\nu) = \emptyset$. It is immediate that $\bigcup_{\nu<\omega_1} \mathfrak{F}_\nu$ is a separation of Y, which is a contradiction. QED(CLAIM)

By the claim, let $\{f_\alpha\}_{\alpha<\omega_1}$ be as in 4.2 now.

By induction, we define sets H_γ, K_γ, $\gamma < \omega_1$, so that:

(i) $H_\gamma, K_\gamma \subseteq \omega_1$;

(ii) $\gamma < \delta \longrightarrow H_\gamma \subseteq H_\delta$ & $K_\gamma \subseteq K_\delta$;

(iii) $H_\gamma \cap K_\gamma = \emptyset$;

(iv) $\gamma \subseteq H_\gamma \cup K_\gamma$.

In order to carry out the definition we need a partial function ν from ω_1 to ω_1 defined by letting $\nu(\alpha)$ be the least element of

$$(\text{closure } W(f_\alpha,\alpha)) \cap (\omega_1 - \alpha) ,$$

if there is such an element, with $\nu(\alpha)$ undefined otherwise.

To commence we set $H_0 = K_0 = \emptyset$. And if γ is a limit ordinal we set

$$H_\gamma = \bigcup_{\nu<\gamma} H_\nu ,$$

$$K_\gamma = \bigcup_{\nu<\gamma} K_\nu .$$

Suppose now that $\gamma = \alpha + 1$ and we have defined H_α, K_α . There are two cases to consider.

Case 1. $f_\alpha : \alpha \to \omega$, $\nu(\alpha)$ is defined, and for all $\beta < \alpha$, if $\nu(\beta)$ is defined then $\nu(\beta) < \alpha$.

Now, $\alpha \subseteq H_\alpha \cup K_\alpha$, so

$$W(f_\alpha,\alpha) = W(H_\alpha,f_\alpha,\alpha) \cup W(K_\alpha,f_\alpha,\alpha) .$$

Hence,

$$\text{closure } W(f_\alpha,\alpha) = \text{closure } W(H_\alpha,f_\alpha,\alpha) \cup \text{closure } W(K_\alpha,f_\alpha,\alpha).$$

Since $\nu(\alpha)$ is defined, either $\nu(\alpha) \in$ <u>closure</u> $W(H_\alpha, f_\alpha, \alpha)$ or $\nu(\alpha) \in$ <u>closure</u> $W(K_\alpha, f_\alpha, \alpha)$ (or both). If $\nu(\alpha) \in$ <u>closure</u> $W(H_\alpha, f_\alpha, \alpha)$, we set

$$K_\gamma = K_\alpha \cup \{\nu(\alpha)\} ;$$

$$H = H_\alpha \cup [\gamma - K_\gamma] .$$

On the other hand, if $\nu(\alpha) \notin$ <u>closure</u> $W(H_\alpha, f_\alpha, \alpha)$ (in which case we must have $\nu(\alpha) \in$ <u>closure</u> $W(K_\alpha, f_\alpha, \alpha)$), we set

$$H_\gamma = H_\alpha \cup \{\nu(\alpha)\} ;$$

$$K_\gamma = K_\alpha \cup [\gamma - H_\gamma].$$

<u>Case 2</u>. Otherwise.

Set
$$H_\gamma = H_\alpha \ ;$$

$$K_\gamma = K_\alpha \cup [\gamma - H_\gamma] .$$

That completes the definition. The only point we need to check is that $H_\gamma \cap K_\gamma = \emptyset$ for all γ. Now, the only thing which could conceivably go wrong here is that for distinct α, α' in Case 1, $\nu(\alpha) = \nu(\alpha')$ and we put (say) $\nu(\alpha)$ into $H_{\alpha+1}$ and $\nu(\alpha')$ into $K_{\alpha'+1}$. But look, if $\alpha < \alpha'$, then for α' to be in Case 1 we must have $(\forall\beta<\alpha)$[if $\nu(\beta)$ is defined then $\nu(\beta) < \alpha'$], so in particular $\nu(\alpha) < \alpha'$. But by definition, $\nu(\alpha') \geqslant \alpha'$, so we must have $\nu(\alpha) \neq \nu(\alpha')$. In other words, nothing can go wrong.

Set
$$H = \bigcup_{\nu<\omega_1} H_\nu \quad , \quad K = \bigcup_{\nu<\omega_1} K_\nu .$$

Clearly, H and K are disjoint. Being subsets of ω_1, they are discrete, and hence closed. By normality we can find disjoint open sets U, V such that $H \subseteq U$ and $K \subseteq V$.

Pick some function $f: \omega_1 \longrightarrow \omega$ so that

$$\alpha \in H \longrightarrow N_{f(\alpha)}(\alpha) \subseteq U ;$$

$$\alpha \in K \longrightarrow N_{f(\alpha)}(\alpha) \subseteq V .$$

Since $H \cup K = \omega_1$, this definition is sound. Let

$$E = \{\alpha \in \omega_1 \mid f \upharpoonright \alpha = f_\alpha \} .$$

By assumption, E is a stationary subset of A_f.

Let $C = \{\alpha \in \omega_1 \mid \alpha$ is a limit ordinal & $(\forall \beta < \alpha)[$ if $\nu(\beta)$ is defined then $\nu(\beta) < \alpha]\}$.

It is not hard to prove that C is closed and unbounded in ω_1. Hence there is a point $\alpha \in E \cap C$. Since $\alpha \in A_f$, (<u>closure</u> $W(f,\alpha)) \cap \omega_1 \neq \alpha$. So as $\alpha \in E$, (<u>closure</u> $W(f_\alpha,\alpha)) \cap \omega_1 \neq \alpha$. Hence $\nu(\alpha)$ is defined. Thus, using now the fact that $\alpha \in C$, we see that Case 1 applied in the definition of $H_{\alpha+1}$ and $K_{\alpha+1}$. This means that either $\nu(\alpha) \in$ (<u>closure</u> $W(H_\alpha,f_\alpha,\alpha)) \cap K_{\alpha+1}$ or else $\nu(\alpha) \in$ (<u>closure</u> $W(K_\alpha,f_\alpha,\alpha)) \cap H_{\alpha+1}$. Hence either $\nu(\alpha) \in$ (<u>closure</u> $W(H,f,\omega_1)) \cap K$ or else $\nu(\alpha) \in$ (<u>closure</u> $W(K,f,\omega_1)) \cap H$. But $W(H,f,\omega_1) \subseteq U$ and $W(K,f,\omega_1) \subseteq V$. Hence either $\nu(\alpha) \in$ (<u>closure</u> $U) \cap K$ or else $\nu(\alpha) \in$ (<u>closure</u> $V) \cap H$. But $H \subseteq U$ and $K \subseteq V$, $U \cap V = \emptyset$, and U and V are open. Thus (<u>closure</u> $U) \cap K$ = (<u>closure</u> $V) \cap H = \emptyset$. We have thus arrived at a contradiction. The theorem is proved. □

5. Further Remarks

The three applications of V = L which we have just given all made use of combinatorial enumeration principles of a similar kind. These principles enabled us to carry through an inductive construction which would normally break down. There are other applications of V = L where it is not possible to isolate a simple combinatorial principle such as $\diamondsuit$ from the proof, but in almost every case V = L is used to help carry through a complicated induction. In the next chapter we give some indication of just why V = L works in this manner.

We have already mentioned that the principle $\diamondsuit$, the simplest of the three principles we have made use of, does not follow from GCH. But the nature of each of our proofs might lead the reader to suspect that GCH would be enough to solve our three problems. (Although there is no intuitive reason for taking GCH as an axiom of set theory.) In which case it would not be that a solution of the problems needs a detailed analysis of what constitutes a set, but rather that more information about the arithmetic of infinite cardinal numbers is required. This is not the case. Using highly complex varieties of the type of procedure outlined in Appendix II, R. B. Jensen has shown that the GCH does not solve Souslin's Problem; S. Shelah has shown that the GCH solves neither the Whitehead Problem nor the CWH problem.

And a final remark concerning the origin of our three results. The original solution to the Souslin Problem using V = L is due to R. B. Jensen, and indeed it is this result which constituted the major breakthrough leading to applications of the Axiom of Constructibility outside set theory. The version of the proof we gave here is due to K. Kunen. The solution of the Whitehead Problem using V = L is due to S. Shelah. Our account owes much to an account by P. Eklof. The result on CWH spaces is due to W. Fleissner.

Chapter V A PROBLEM IN MEASURE THEORY

This chapter is intended to serve two purposes. On the one hand we shall prove a result in measure theory. In order to obtain our proof we shall have need to investigate the notion of a constructible set a little more thoroughly than before, and thereby we shall achieve our second goal — to give some indication of how the Axiom of Constructibility works.

1. Extensions of Lebesgue Measure.

By a **measure** on a σ-algebra, $\mathcal{R}$, of subsets of a set X we mean a countably additive function μ from $\mathcal{R}$ into the closed unit interval $[0,1]$ which vanishes on the empty set and on any points in $\mathcal{R}$, and which takes the value 1 on the whole space, X. (This definition excludes Lebesgue measure on $\mathbb{R}$, of course, but includes Lebesgue measure on the unit interval $[0,1]$, so there is no great loss.)

Our starting point is the following question. Is there an extension of Lebesgue measure on $[0,1]$ to a measure μ defined on all subsets of $[0,1]$? The "usual" proof that there is a non Lebesgue measurable subset of $[0,1]$ shows that such an extension cannot be translation invariant.

Whilst we are on the subject, let us remark that in Lebesgue measurability on $[0,1]$ we have a further example of a necessary use of the Axiom of Choice. It is known from work of Solovay that the assumption of the Axiom of Choice is strictly necessary in order to construct a set of reals which is not Lebesgue measurable. So why did we not include the measure extension problem in Chapter I ? Well, because its status is not quite the same as the problems considered there, and moreover, in order to investigate the problem we need to develop the theory of constructibility to a degree not required for the other problems.

A classical result of Ulam tells us that if the CH holds, then there is no extension of Lebesgue measure on [0,1] to a measure defined on all subsets of this interval. (We prove this fact below.) Hence the Axiom of Constructibility certainly answers our basic question. But it immediately raises a new problem. Is there any uncountable set X for which there is a measure μ defined on all subsets of X ? Depending on your viewpoint, this question is either of fundamental importance in measure theory, or of peripheral interest in measure theory. But either way it is of interest! It is known that a positive solution is impossible in Zermelo-Fraenkel set theory with just the Axiom of Choice. (We shall see why in a little while.) However, it is certainly possible that one could obtain a negative solution from these assumptions. To date no one has been able to do this. What we can (and shall) do is obtain a negative solution from the Axiom of Constructibility.

2. The Measure Problem

We investigate the problem: does there exist an uncountable set X and a measure μ defined on all subsets of X ?

First let us observe that it is enough to consider cardinal numbers κ instead of arbitrary sets X. For, given any pair X, μ as required, if we let $\kappa = |X|$ and if $f: X \leftrightarrow \kappa$, then f induces from μ in the obvious way a measure defined on all subsets of κ . By a measure on a cardinal κ we shall henceforth mean a measure defined on all subsets of κ .

Let μ be a measure on an uncountable cardinal κ . Let λ be any uncountable cardinal. We shall say that μ is λ-additive iff, whenever A_ν, $\nu < \lambda'$, is a collection of fewer than λ sets of measure zero, then $\bigcup_{\nu<\lambda} A_\nu$ has measure zero. Thus, by definition, any measure is $\aleph_1$-additive.

Suppose μ is a measure on the uncountable cardinal κ . There is clearly a largest cardinal λ such that μ is λ-additive. By our remark above, $\lambda \geq \aleph_1$. And since κ is the union of its one-element subsets, $\lambda \leq \kappa$. By definition of λ, there is a set $A \subseteq \kappa$ of positive measure which is the union of λ disjoint sets of measure zero:

$$A = \bigcup_{\nu<\lambda} A_\nu .$$

Define a map $f: A \rightarrow \lambda$ by setting

$$f(a) = \nu \quad \text{iff} \quad a \in A_\nu .$$

For $B \subseteq \lambda$ now, set

$$\sigma(B) = \frac{\mu(f^{-1}[B])}{\mu(A)} .$$

It is easily seen that σ is a λ-additive measure on λ. Hence, defining a measure μ on a cardinal κ to be <u>strong</u> if it is κ-additive, we see that we have proved the following lemma.

2.1 <u>Lemma</u>

Suppose that for some uncountable cardinal κ there is a measure defined on all subsets of κ. Then there is an uncountable cardinal $\lambda \leq \kappa$ such that there is a strong measure defined on all subsets of λ. □

Thus, in order to show that on no uncountable cardinal is there a measure (defined on all sets), it suffices to show that on no uncountable cardinal is there a strong measure. As a first step we prove that if there is a strong measure on the uncountable cardinal κ, then κ is an <u>inaccessible cardinal</u>.

So what is an inaccessible cardinal ? Well, we know already what it means for a cardinal to be <u>regular</u>. And we also know what it means for a cardinal to be a <u>limit cardinal</u>. The first infinite cardinal, $\aleph_0$, is both regular and a limit cardinal. But are there any other cardinals which are both regular and a limit cardinal ? It is a consequence of the Gödel Incompleteness Theorem that a positive answer to this question is not possible. But considerations outside the scope of this book lead to the conclusion that a negative answer is extremely unlikely (indeed, as unlikely as the possibility that Zermelo-Fraenkel Set Theory is inconsistent). It is also known that if a negative solution were to be obtained assuming the Axiom of Constructibility, then a negative answer would already follow from Zermelo-Fraenkel Set Theory with just the Axiom of Choice. Roughly speaking, an inaccessible cardinal is "very large", and certainly much bigger than $\aleph_1, \aleph_2, \ldots, \aleph_n, \ldots, \aleph_\omega, \ldots, \aleph_{\omega_1}, \ldots$ etc.

Fix μ a strong measure on an uncountable cardinal κ from now on.

2.2 Lemma

If $\xi < \kappa$, then $\{\alpha \mid \alpha < \xi\}$ has measure zero.

Proof: Since μ is strong. □

2.3 Lemma

κ is regular.

Proof: Suppose not. Then there is a $\lambda < \kappa$ and ordinals $\kappa_\nu < \kappa$, $\nu < \lambda$, such that

$$\kappa = \bigcup_{\nu<\lambda} \kappa_\nu \, .$$

By 2.2, $\mu(\kappa_\nu) = 0$ for all ν . So as μ is strong, we get $\mu(\kappa) = 0$, a contradiction. □

The proof that κ is a limit cardinal will take a little longer.

We say that μ is normal iff, whenever $B \subseteq \kappa$ has positive measure and $f : B \to \kappa$ is such that $f(\xi) < \xi$ for all $\xi \in B$, there is a $B' \subseteq B$ of positive measure such that f is constant on B'. As a first step in our proof we show that we may assume that our measure μ is normal. We require some auxiliary definitions.

Let $f : A \to \kappa$, where $A \subseteq \kappa$. We say that f is almost bounded if there is a $\lambda < \kappa$ such that $\{\xi \in A \mid f(\xi) > \lambda\}$ has measure zero. We say that f is nowhere bounded if for each $\lambda < \kappa$, $\{\xi \in A \mid f(\xi) \leq \lambda\}$ has measure zero. We say that f is incompressible if f is nowhere bounded and whenever $B \subseteq A$ has positive measure and $g : B \to \kappa$ and $g(\xi) < f(\xi)$ for all $\xi \in B$, then g is almost bounded. These concepts are only important if A has positive measure, of course, for if A has measure zero, then any map f is simultaneously almost bounded, nowhere bounded, and incompressible.

2.4 Lemma

Let $f:A \to \kappa$ be nowhere bounded. Then we can write A as the disjoint union of sets B and C such that:

(1) $f \upharpoonright B$ is incompressible;

(2) there is a $g:C \to \kappa$ such that $g(\xi) < f(\xi)$ for all $\xi \in C$ and such that g is nowhere bounded.

Proof: Using Zorn's lemma we obtain a maximal family

$$\mathcal{F} = \{(C_i, g_i) \mid i \in K\}$$

such that:

(1) $C_i \subseteq A$ has positive measure;

(2) $g_i:C_i \to \kappa$ is nowhere bounded;

(3) $\xi \in C_i \to g_i(\xi) < f(\xi)$;

(4) if i,j are distinct elements of K, then $C_i \cap C_j = \emptyset$.

By (1) and (4), K must be countable. (If K were uncountable, then for some positive number n, C_i would have measure greater than $1/n$ for uncountably many i, contrary to the measure being finite.) We set

$$C = \bigcup_{i \in K} C_i \quad , \quad g = \bigcup_{i \in K} g_i .$$

Since K is countable, g is nowhere bounded. And clearly, $\xi \in C \to g(\xi) < f(\xi)$. Set $B = A - C$. If B has measure zero, $f \upharpoonright B$ is trivially incompressible. If B has positive measure, the maximality of $\mathcal{F}$ implies that $f \upharpoonright B$ is incompressible. □

2.5 Lemma

There is an incompressible function $f: \kappa \to \kappa$.

Proof: We define a sequence of sets $\{A_n\}$ and functions $h_n:A_n \to \kappa$ by induction on n. To commence we set $A_0 = \kappa$ and let h_0 be the identity function on κ. (By 2.2, h_0 is nowhere bounded.) Suppose now that $n = k + 1$ and that we have defined A_k and h_k with $h_k:A_k \to \kappa$ nowhere bounded. We apply 2.4 to A_k, h_k to obtain a set $A_{k+1} \subseteq A_k$ and a map

$h_{k+1}:A_{k+1}\to\kappa$ which is nowhere constant, such that $h_{k+1}(\xi)<h_k(\xi)$ for $\xi\in A_{k+1}$ and $h_k\restriction A_k-A_{k+1}$ is incompressible.

We show that $\bigcap_{n=1}^{\infty}A_n=\emptyset$. Suppose otherwise, and let ξ lie in this intersection. Then $h_0(\xi)>h_1(\xi)>\dots$, so $\{h_n(\xi)\}$ is a strictly decreasing sequence of ordinals. But the ordinals are well-ordered, so this is impossible.

By the above,

$$\kappa = \bigcup_{n=1}^{\infty}(A_n - A_{n+1})$$

is a disjoint union and we may define $h:\kappa\to\kappa$ by

$$h(\xi) = h_n(\xi) \text{ iff } \xi\in A_n - A_{n+1}.$$

It is easily checked that h is incompressible. □

Let $h:\kappa\to\kappa$ be incompressible now. Define a function $\nu:\mathcal{P}(\kappa)\to[0,1]$ by

$$\nu(A) = \mu(h^{-1}[A]).$$

It is easily checked that ν is a strong measure on κ . We prove that ν is normal. Suppose A has positive ν-measure and that $g:A\to\kappa$ is such that $g(\xi)<\xi$ for all $\xi\in A$. Let $B=h^{-1}[A]$. Then B has positive μ-measure. Let $f=g\circ h$. Thus $f:B\to\kappa$. If $\gamma\in B$, $f(\gamma)=g(h(\gamma))<h(\gamma)$. So as h is incompressible there is a $\lambda<\kappa$ for which $\{\gamma\in B\,|\,f(\gamma)\le\lambda\}$ has positive μ-measure. Since μ is strong and $\lambda<\kappa$ there is a $\lambda'\le\lambda$ such that $D=\{\gamma\in B\,|\,f(\gamma)=\lambda'\}$ has positive μ-measure. Let $E=\{\gamma\in A\,|\,g(\gamma)=\lambda'\}$. Then $\gamma\in D\leftrightarrow g(h(\gamma))=\lambda'\leftrightarrow h(\gamma)\in E$. Thus $D=h^{-1}[E]$. Hence E has positive ν-measure. But $E\subseteq A$ and g is constant on E, so we are done.

We thus see that without loss of generality we may assume that in fact the measure μ is normal.

2.6 Lemma

Let A have positive measure. Let $h:A\to\kappa$ be such that $h(\xi)<\xi$ for all $\xi\in A$. Then h is almost bounded.

Proof: Let $E=\{\lambda<\kappa\,|\,h^{-1}[\{\lambda\}]$ has positive measure$\}$.

Clearly, E must be countable.

Let $$B = \{\gamma \in A \mid h(\gamma) \notin E\}.$$
B must have measure zero. For otherwise, by normality, there is a $\lambda < \kappa$ and a $B' \subseteq B$ of positive measure such that $h(\gamma) = \lambda$ for $\gamma \in B'$, giving $\lambda \in E$, contrary to the definition of B.

Since κ is regular, $\lambda_0 = \sup(E) < \kappa$. But
$$\{\gamma \in A \mid h(\gamma) > \lambda_0\} \subseteq B.$$
Thus h is almost bounded. □

2.7 <u>Lemma</u>

For almost all $\alpha \in \kappa$, α is a regular cardinal.

Proof: Suppose not. Let $E = \{\alpha \mid cf(\alpha) < \alpha\}$. Thus E has positive measure. Hence, by normality, there is a $\lambda < \kappa$ such that $E_1 = \{\alpha \mid cf(\alpha) = \lambda\}$ has positive measure. For each $\alpha \in E_1$, pick a mapping $h_\alpha : \lambda \to \alpha$ such that $\sup(h_\alpha[\lambda]) = \alpha$. Define for each $\xi < \lambda$ a map $g_\xi : E_1 \to \kappa$ by $g_\xi(\alpha) = h_\alpha(\xi)$. Then for all $\alpha \in E_1$, $g_\xi(\alpha) = h_\alpha(\xi) < \alpha$. Applying 2.6 to g_ξ, we see that there is a set N_ξ of measure zero and an ordinal $\gamma_\xi < \kappa$ such that $g_\xi(\alpha) \leq \gamma_\xi$ if $\alpha \in E_1 - N_\xi$. Set $\gamma = \sup\{\gamma_\xi \mid \xi < \lambda\}$. Since κ is regular, $\gamma < \kappa$. Put $E_2 = E_1 - \bigcup_{\xi<\lambda} N_\xi$. Since μ is strong, $\mu(E_2) > 0$. Now, for $\alpha \in E_2$,
$$\alpha = \sup\{g_\xi(\alpha) \mid \xi < \lambda\} \leq \sup\{\gamma_\xi \mid \xi < \lambda\} \leq \gamma.$$
Thus $E_2 \subseteq \{\alpha \mid \alpha < \gamma\}$. Since $\mu(E_2) > 0$, this contradicts 2.2. □

2.8 <u>Theorem</u>

κ is inaccessible.

Proof: We know that κ is regular. So if κ is not inaccessible, there must be a cardinal λ such that $\kappa = \lambda^+$, the least cardinal greater than λ. Then
$$\{\alpha \mid \alpha \text{ is regular}\} \subseteq \{\alpha \mid \alpha \leq \lambda\},$$
so by 2.2 $\{\alpha \mid \alpha$ is regular$\}$ has measure zero, contrary to 2.7. □

Notice that since any inaccessible cardinal is greater than $\aleph_1$, 2.8 provides us with a proof of Ulam's theorem that, assuming CH, there is no extension of Lebesgue measure to a measure defined on all sets of reals.

3. A Theorem in Model Theory[1]

A predicate language consists of the following:

logical connectives $\wedge$ (and), $\vee$ (or), $\neg$ (not), $\rightarrow$ (implies);

quantifiers $\forall$ (for all), $\exists$ (there exists);

variables $v_0, v_1, v_2, \ldots$;

brackets (,) ;

(possibly) n-ary predicate symbols P^n_1, P^n_2, ... (each n).

The formulas of the language are built up as follows:

1. If $x_1, \ldots, x_n$ are variables, then $P^n_m(x_1, \ldots, x_n)$ is a formula;
2. If φ, ψ are formulas, so are $(\varphi \wedge \psi)$, $(\varphi \vee \psi)$, $(\neg \varphi)$, $(\varphi \rightarrow \psi)$;
3. If φ is a formula, then so are $(\forall v_n \varphi)$, $(\exists v_n \varphi)$.

The meanings of the above operations are self-evident. If a variable v_n occurs in a formula φ within the scope of a quantifier $\forall v_n$ or $\exists v_n$, we say that occurrence of v_n in φ is bound; otherwise the occurrence is free. (Thus a free variable is one for which we could "substitute" any value we chose from the values available. And a bound variable is one which is an integral part of the meaning of the formula in which it occurs.)

Let S be any predicate language. An S-structure is a structure of the form $\mathfrak{A} = (A, R^1_1, R^1_2, \ldots, R^2_1, R^2_2, \ldots)$. If φ is a formula of S and if $x_1, \ldots, x_m$ are the free variables of φ, and if $a_1, \ldots, a_m$ are elements of A, we say that φ is satisfied in $\mathfrak{A}$ at the point $(a_1, \ldots, a_m)$ iff φ is a true assertion about the structure $\mathfrak{A}$ when R^n_i interprets P^n_i and x_i denotes a_i ; we write

$$\mathfrak{A} \models \varphi \; [a_1, \ldots, a_m]$$

in this case. (The notion of satisfaction can be rigourously defined , but for our purposes the obvious intuitive meaning is adequate.)

---

1. Although this section is, in a strict sense, self-contained, it may require some prior knowledge of the material for a proper understanding. The material covered is also required for section 4.

For example, let S be the predicate language for group theory. This has only one binary predicate, denoting equality of two group elements, and one ternary predicate denoting x.y = z. Any group will be an S-structure. The fact that a group $\underset{\sim}{G}$ is abelian can be expressed in any of the following forms:

(i) $\underset{\sim}{G} \models \forall x \forall y \forall z(x.y = z \rightarrow y.x = z)$

(ii) for all g,h in $\underset{\sim}{G}$, $\underset{\sim}{G} \models \forall z(x.y = z \rightarrow y.x = z) \; [g,h]$

(iii) for all g,h,k in $\underset{\sim}{G}$, $\underset{\sim}{G} \models (x.y = z \rightarrow y.x = z) \; [g,h,k]$.

(For reasons of clarity, one rarely retains the formalism of the definitions for very long. Thus, in the above we write x.y = z instead of P(x,y,z) for the appropriate P. Likewise we use x,y,z etc. instead of $v_0, v_1, v_2, \ldots$.)

If we have two S-structures $\mathfrak{A}$, $\mathfrak{B}$, we say that $\mathfrak{A}$ is an elementary substructure of $\mathfrak{B}$, and write $\mathfrak{A} \prec \mathfrak{B}$, if $\mathfrak{A}$ is a substructure of $\mathfrak{B}$ in the obvious sense, and for all formulas φ of S with free variables (say) $x_1,\ldots,x_n$ and all elements $a_1,\ldots,a_n$ of $\mathfrak{A}$,

$$\mathfrak{A} \models \varphi \; [a_1,\ldots,a_n] \quad \text{iff} \quad \mathfrak{B} \models \varphi \; [a_1,\ldots,a_n] \; .$$

(Loosely speaking, $\mathfrak{A} \prec \mathfrak{B}$ means that although $\mathfrak{B}$ contains more elements than $\mathfrak{A}$, $\mathfrak{A}$ and $\mathfrak{B}$ looke very much alike with regards to their internal structure. For example, if $\underset{\sim}{G}$ and $\underset{\sim}{H}$ are groups and $\underset{\sim}{G} \prec \underset{\sim}{H}$, then $\underset{\sim}{G}$ will be abelian iff $\underset{\sim}{H}$ is abelian.)

The notion of elementary substructure was introduced by Tarski , who also showed that the construction of elementary substructures can be reduced to a purely algebraic construction. More precisely:

3.1 Lemma

Let S be a predicate language. Let $\mathfrak{A}$ be an S-structure, $\mathfrak{A} = (A,\ldots)$. Then there are finitary functions $f_1, f_2, \ldots$ defined on A into A (called Skolem functions) such that for any set $X \subseteq A$, if $B = \bigcup_{n=1}^{\infty} f_n[X]$ and if $\mathfrak{B}$ is the structure $\mathfrak{A}$ relativised to B, then $X \subseteq \mathfrak{B} \prec \mathfrak{A}$. □

A proof of 3.1 can be found in any elementary text on mathematical logic. As an illustration of the use of 3.1 we prove the following classical theorem of model theory.

3.2 Theorem (The Löwenheim-Skolem-Tarski Theorem)

Let S be a predicate language, $\mathfrak{A}$ an S-structure of infinite cardinality κ. Let $\lambda \leqslant \kappa$ be an infinite cardinal, and let X be a subset of $\mathfrak{A}$ of cardinality λ. Then there is a $\mathfrak{B} \prec \mathfrak{A}$ such that $X \subseteq \mathfrak{B}$ and $|\mathfrak{B}| = \lambda$.

Proof: Let $\{f_n\}$ be a sequence of skolem functions for $\mathfrak{A}$. Let $B = \bigcup_{n=1}^{\infty} f_n[X]$. By 3.1, if $\mathfrak{B}$ is the relativisation of $\mathfrak{A}$ to B, then $\mathfrak{B} \prec \mathfrak{A}$. Since we clearly have $|B| = \lambda$, we are done. □

Using 3.1, we shall also prove the following rather specialised version of the above theorem. This will be instrumental in our solving the measure problem from the Axiom of Constructibility.

3.3 Theorem

Let κ be an uncountable cardinal which carries a strong measure. Let S be any predicate language which contains a unary predicate symbol U, and let

$$\mathfrak{A} = (A, U, \dots)$$

be any S-structure such that $|A| \geqslant \kappa$ and $|U| < \kappa$. Then there is a $\mathfrak{B} \prec \mathfrak{A}$,

$$\mathfrak{B} = (B, U \cap B, \dots),$$

such that $|B| = \kappa$ and $|U \cap B| \leqslant \aleph_0$. □

The proof will take some time. Let us remark that the difficulty is to keep B large whilst ensuring that $U \cap B$ is small. In order to do this we must rely heavily upon the fact that κ carries a strong measure.

To commence the proof, notice that we may assume that A is itself a cardinal number (just transfer the structure from A to its cardinal number), and hence that $\kappa \subseteq A$. We fix some sequence $\{f_n\}$ of Skolem functions for $\mathfrak{A}$ in accordance with 3.1. Suppose f_n is k(n)-ary. We define a k(n)-ary function $\bar{f}_n$ from $\kappa^{k(n)}$ to U by:

$$\bar{f}_n(x_1,\ldots,x_{k(n)}) = \begin{cases} f_n(x_1,\ldots,x_{k(n)}), & \text{if this is an element of } U, \\ \text{an arbitrary element of } U, & \text{otherwise.} \end{cases}$$

In order to prove 3.3 it clearly suffices to find a subset X of κ of cardinality κ such that $|\bar{f}_n[X]| \leq \aleph_0$ for all n. (For then the range of the functions f_n on X will form the domain of the required substructure.) Now, by introducing extra functions $\bar{f}_n$ if necessary, we may assume that all the functions $\bar{f}_n$ are commutative (i.e. the order in which the arguments appear is irrelevant), and that if $x_i = x_j$ for some i,j with $1 \leq i < j \leq k(n)$, then for some m with $k(m) = k(n) - 1$,

$$\bar{f}_n(x_1,\ldots,x_{k(n)}) = \bar{f}_m(x_1,\ldots,x_{j-1},x_{j+1},\ldots,x_{k(n)}).$$

We are thus reduced to proving the following. If $X \subseteq \kappa$, let $X^{[n]}$ denote the set of all strictly increasing n-tuples from X. Let $\{f_n\}$ be a sequence of functions

$$f_n: \kappa^{[k(n)]} \to \lambda,$$

where $\lambda < \kappa$. Then there is a subset X of κ of cardinality κ such that

$$|f_n[X^{[k(n)]}]| \leq \aleph_0 \text{ for all n.}$$

(Notice that we have, in effect, replaced the set U by its cardinality, λ, for convenience.) Since the intersection of countably many sets of measure one has measure one (and thus has cardinality κ), we see finally that it suffices to prove the following:

3.4 <u>Lemma</u>

Let κ be an uncountable cardinal and let μ be a normal strong measure on κ. Let $f: \kappa^{[n]} \to \lambda < \kappa$. Then there is a set $D \subseteq \kappa$ of measure one such that

$$|f[D^{[n]}]| \leq \aleph_0.$$

Proof: By induction on n.

<u>Case 1</u>: n = 1. Let $E = \{\xi \mid f^{-1}[\{\xi\}]$ has positive measure $\}$. Clearly, E is countable. Let $N = \{\gamma \mid f(\gamma) \notin E\}$. Since $f(\xi) < \lambda < \kappa$ for all ξ, the normality of μ implies that $\mu(N) = 0$. Thus $D = \kappa - N$ suffices.

<u>Case 2</u>: n = k+1. For each α, define a map $h_\alpha: \kappa^{[k]} \to \lambda$ by:

$$h_\alpha(x_1,\ldots,x_k) = \begin{cases} f(\alpha,x_1,\ldots,x_k), & \text{if } \alpha < x_1, \\ 0, & \text{otherwise.} \end{cases}$$

By the induction hypothesis we can find a set D_α of measure one such that $D_\alpha \subseteq \kappa - \alpha$ and $|h_\alpha[D_\alpha^{[k]}]| \leq \aleph_0$. Set $S_\alpha = h_\alpha[D_\alpha^{[k]}]$. Thus $|S_\alpha| \leq \aleph_0$. Let $s_\alpha : \omega \xrightarrow{\text{onto}} S_\alpha$. For each n, define $g_n : \kappa \to \lambda$ by

$$g_n(\alpha) = s_\alpha(n).$$

By case 1 there is a set N_n of measure zero and a countable set $E_n \subseteq \kappa$ such that

$$\alpha \in \kappa - N_n \to g_n(\alpha) \in E_n.$$

Set $\quad F = \kappa - \bigcup_{n=0}^{\infty} N_n \quad , \quad E = \bigcup_{n=0}^{\infty} E_n .$

Then $\mu(F) = 1$ and E is countable. And for all n,

$$\alpha \in F \to g_n(\alpha) \in E.$$

Thus,

$$\alpha \in F \to S_\alpha = \{s_\alpha(n) | n \in \omega\} = \{g_n(\alpha) | n \in \omega\} \subseteq E.$$

Now set

$$D = \{\gamma \in F \mid \gamma \in \bigcap_{\alpha < \gamma} D_\alpha\}.$$

We prove that $f[D^{[n]}] \subseteq E$. Let $x \in D^{[n]}$ and let $\alpha = \min(x)$, and let $y \in D^{[k]}$ be such that $x = \{\alpha\} \cup y$. By definition of D, $y \in D_\alpha^{[k]}$. Thus, $f(x) = h_\alpha(y) \in S_\alpha$. But $\alpha \in D \subseteq F$, so $S_\alpha \subseteq E$. Hence $f(x) \in E$. Thus $f[D^{[n]}] \subseteq E$.

It thus suffices to show that $\mu(D) = 1$. Well suppose not. Then

$$D' = \{\gamma \in F \mid \gamma \notin \bigcap_{\alpha < \gamma} D_\alpha\}$$

has positive measure. Define $g : D' \to \kappa$ by

$$g(\gamma) = \text{the least } \alpha < \gamma \text{ such that } \gamma \notin D_\alpha.$$

Then $g(\gamma) < \gamma$ for all $\gamma \in D'$, so by normality there is a set $D'' \subseteq D'$ of positive measure such that g is constant on D'', say with value α_0. Then,

$$\gamma \in D'' \to \gamma \notin D_{\alpha_0}.$$

But D_{α_0} has measure one, so this is impossible. The proof is complete. □

3.4 completes the proof of 3.3.

4. The Condensation Lemma

The most significant property of the constructible hierarchy, L_α, $\alpha \in On$, is the condensation property.

4.1 Theorem (Condensation Lemma)

Let κ be an infinite cardinal. Suppose $(N, \in, =) \prec (L_\kappa, \in, =)$. Then there is a unique ordinal γ and a unique structural isomorphism π such that

$$\pi : (N, \in, =) \cong (L_\gamma, \in, =).$$

Moreover, if β is an ordinal such that $\{\alpha \mid \alpha < \beta\} \subseteq N$, then for all $\alpha < \beta$, $\pi(\alpha) = \alpha$, and for all $x \in N$ such that $x \subseteq \beta$, $\pi(x) = x$. □

The proof of 4.1 is outside the scope of this book. It can be found in [De]. Coupled with 3.2, 4.1 tells us that no matter how large the cardinal κ is, there is always a countable ordinal γ such that the structure $(L_\gamma, \in, =)$ looks just like $(L_\kappa, \in, =)$. Now, the Axiom of Constructibility says that the universe is just the union of all the partial universes L_κ. Hence, assuming V = L provides us with an extremely uniform universe of sets; if anything exciting happens "high up" in the universe, it is already mirrored within some countable partial universe L_γ. This is why it is often possible to carry out constructions by transfinite induction using the Axiom of Constructibility. Indeed, the combinatorial principle $\diamondsuit$ introduced in IV.1 is a relatively easy, direct consequence of the condensation lemma. (One defines the $\diamondsuit$ sequence by a transfinite induction, using the condensation lemma to prove that the sequence does what we require of it. Any result proved using $\diamondsuit$ could be proved by a direct appeal to the condensation lemma itself. However, this usually requires some competence in techniques of Mathematical Logic, so in practice the non-specialist will probably find it easier to apply $\diamondsuit$ as we did in, say, IV.2. For a similar reason we do not give here the derivation of $\diamondsuit$. It can be found in [De].)

Using 4.1, we can obtain a very quick proof of the GCH from V = L. First we need a lemma which is itself rather illuminating. We have already mentioned that as we proceed up the constructible hierarchy from, say, stage ω, new subsets of ω will continue to appear (which is not the case with the cumulative hierarchy, of course). The following lemma tells us that this ceases as soon as it possibly can.

4.2 Lemma

Assume V = L. Let κ be an infinite cardinal. Let $x \subseteq \kappa$. Then $x \in L_{\kappa^+}$.

Proof: Since $V = \bigcup_\alpha L_\alpha$, we can find a cardinal λ such that $x \in L_\lambda$. By 3.2 we can find an elementary submodel

$$(N, \in, =) \prec (L_\lambda, \in, =)$$

such that $\{\alpha | \alpha < \kappa\} \cup \{x\} \subseteq N$ and $|N| = \kappa$. By 4.1, let

$$\pi : (N, \in, =) \cong (L_\gamma, \in, =).$$

Now, $|L_\gamma| = |N| = \kappa$. But it is easily proved (by transfinite induction) that for any infinite ordinal ξ, $|L_\xi| = |\xi|$. Hence $\gamma < \kappa^+$. Hence $L_\gamma \subseteq L_{\kappa^+}$. But $x \in N$ and $x \subseteq \kappa = \{\alpha | \alpha < \kappa\} \subseteq N$, so by 4.1, $\pi(x) = x$. Thus $x \in \pi[N] = L_\gamma$, giving $x \in L_{\kappa^+}$, as required. □

4.3 Theorem

Assume V = L. Then GCH holds.

Proof: Since $|L_\xi| = |\xi|$ for all infinite ξ, and since we know that $2^\kappa = |\mathcal{P}(\kappa)| > \kappa$ for any infinite cardinal κ, it suffices to show that $\mathcal{P}(\kappa) \subseteq L_{\kappa^+}$. But this is immediate by 4.2. □

5. Solution to the Measure Problem.

5.1 Theorem

Assume V = L. Let κ be any uncountable cardinal. Then there is no strong measure on κ.

Proof: Suppose otherwise. Let μ be a normal strong measure on κ.

Now, by 4.2 we have:

$$(L_\kappa, L_{\omega_1}, \in, =) \models \varphi\ [\omega] \quad \dots\dots\dots\dots\dots\ (*),$$

where φ is the formula

$$(\forall x)(x \subseteq y \rightarrow P(x)),$$

and where L_{ω_1} is the unary predicate which interprets P (so P(x) "means" $x \in L_{\omega_1}$), and where ω interprets the free variable y. To see that (*) is valid, we just ask ourselves what φ says when interpreted in the structure $(L_\kappa, L_{\omega_1}, \in, =)$. It says that L_{ω_1} contains all subsets of ω ; which we know to be true by 4.2. (The astute reader may have noticed that our last sentence is not entirely correct. Certainly, 4.2 tells us that φ is a true statement about ω and L_{ω_1}. But why should it then follow that φ is true when interpreted in the partial universe L_κ, which is what (*) actually asserts? Well, the point is that the truth or falsity of φ clearly only depends on ω, L_{ω_1}, and the collection of all subsets of ω. No other sets are involved. But by 4.2, all of these sets are in the partial universe L_κ. Hence φ will be true in L_κ iff it is true.)

By 3.3 there is a set N of cardinality κ such that $|N \cap L_{\omega_1}| \leq \aleph_0$ and

$$(N, L_{\omega_1} \cap N, \in, =) \prec (L_\kappa, L_{\omega_1}, \in, =).$$

Then, clearly, $(N, \in, =) \prec (L_\kappa, \in, =)$, so by 4.1, let

$$\pi : (N, \in, =) \cong (L_\gamma, \in, =).$$

Then, $|\gamma| = |L_\gamma| = |N| = \kappa$. So $\gamma \geq \kappa$. (In fact it can be shown that $\gamma = \kappa$.)

Let $W = \pi[L_{\omega_1} \cap N]$. Thus $|W| \leq \aleph_0$. Clearly,

$$\pi : (N, L_{\omega_1} \cap N, \in, =) \cong (L_\gamma, W, \in, =).$$

Now, (1) $(L_\kappa, L_{\omega_1}, \in, =) \models \varphi[\omega]$.

So, (2) $(N, L_{\omega_1} \cap N, \in, =) \models \varphi[\omega]$.

So, applying π,

(3) $(L_\gamma, W, \in, =) \models \varphi[\pi(\omega)]$.

Now, every positive integer n is definable in the partial universe L_κ by a sentence in the language of set theory. (The sentence which describes explicitly the way in which n is built up starting from the empty set suffices. Admittedly this sentence will be rather long if n is large, but that will not affect matters at all.) Hence, as $(N, \in, =) \prec (L_\kappa, \in, =)$, we see that N will contain all positive integers. Similarly, ω will be an element of N. (In particular, this justifies our deduction of (2) from (1) above.) But by 4.1, we see that this implies that $\pi(\omega) = \omega$. Hence from (3):

(4) $(L_\gamma, W, \in, =) \models \varphi[\omega]$.

What does (4) mean ? It means that in the partial universe L_γ, all subsets of ω lie in the set W. But $\gamma \geq \kappa$, so the partial universe L_γ contains all subsets of ω. Hence (4) implies that W actually does contain all subsets of ω. But W is countable, so this is impossible. The proof is complete. □

By virtue of our discussion in section 1, we see from 5.1 that we have now proved the following result:

5.2 Theorem

Assume V = L. Then for no uncountable set X is there a measure defined on all subsets of X. □

6. Historical Remark

The solution to the Measure Problem as we have presented it here is due to R. M. Solovay. Weaker results had previously been obtained by S. Ulam , D. H. Fremlin, R. B. Jensen, and D. Scott. (In particular, Scott had proved that V = L implies that no uncountable cardinal carries a two-valued measure defined on all subsets. It is possible to deduce the solution to the general measure problem from this result. Both solutions are due to Solovay, the one given here being based upon work of F. Rowbottom.)

Appendix I AXIOMS FOR SET THEORY

In Chapter II we developed the system ZF of set theory, and claimed that this system was adequate for almost all our purposes, at least if we assumed AC as well. But we did not spend much time analysing the definition of the system itself. We just relied upon an intuitive idea of what a set should be, and proceeded from there to construct the Zermelo hierarchy of sets. Since the ZF system remains with us even if we abandon the intuition behind it and opt for constructible set theory, it is worth while analysing just what assumptions the ZF system entails.

When we developed the ZF system, we took the ordinal numbers as basic, so let us continue for the time being to do this. So our first question is: Given the ordinal number system, what assumptions about sets are required in order to construct the Zermelo hierarchy?

Well, we certainly must be able to form the power set of a given set. So let us formulate this assumption as an <u>axiom</u>.

<u>Power Set Axiom</u>

If x is a set, then there is a set consisting of all subsets of x.

This axiom facilitates our passing from V_α to $V_{\alpha+1}$. What about the defintion of V_α when α is a limit ordinal. We then set

$$V_\alpha = \bigcup_{\beta<\alpha} V_\beta .$$

So we certainly must be able to form the union of any set of sets:

<u>Axiom of Union</u>

If x is a set, there is a set whose members are precisely the members of the members of x (this set being denoted by $\bigcup x$).

Then for α a limit ordinal we may set $V_\alpha = \bigcup\{V_\beta \mid \beta<\alpha\}$. But wait a moment.

How do we know that $\{V_\beta \mid \beta < \alpha\}$ is a set? The answer is we do not. Or, to put it another way, we are making an assumption when we take this collection to be a set. (Admittedly it looks a perfectly reasonable assumption, but nonetheless it is an assumption.) We certainly know that $\alpha = \{\beta \mid \beta < \alpha\}$ is a set. (We are assuming the ordinal number system as basic at the moment.) And we obtain the collection $\{V_\beta \mid \beta < \alpha\}$ if we replace each element β of α by V_β. This leads to the Axiom of Replacement. Roughly speaking, this says that if we have a set and we replace each element of this set by some other set, then the new collection is also a set. In order to make this more precise, we must say how the replacement process involved here is to be specified. (i.e. What rules of replacement are allowed?) Unless the original set is finite, we cannot explicitly say which elements are to be replaced by which new sets. We need some general description of the replacement procedure. Since we shall clearly only be interested in replacement procedures of a bona fide mathematical nature, the correct formulation of what is required should now be obvious. As allowable replacement procedures we take those and only those which are describable in our language LST. Formally now:

Axiom of Replacement

If x is a set and F is a function from x to sets which is definable in LST, then $\{F(a) \mid a \in x\}$ is a set.

Since the function F on α defined by $F(\beta) = V_\beta$ is clearly (is this clear?) definable by a formula of LST, the axiom of replacement now tells us that $\{V_\beta \mid \beta < \alpha\}$ $= \{F(\beta) \mid \beta < \alpha\}$ is a set.

Given the ordinals then, the above axioms enable us to define the Zermelo hierarchy of sets. We now ask ourselves what principles are necessary in order to define the ordinals. It turns out that we do not need any more "non-obvious" ones (although some of them are so obvious as to be "non-obvious").

To start off with, we must assume there is a set with no members.

Null Set Axiom

There is a set which has no members.

That provides us with the first ordinal, 0. Next, if α is an ordinal, we shall want to define the next ordinal, $\alpha+1$, as the set $\alpha \cup \{\alpha\}$. This can be written as $\bigcup\{\alpha,\{\alpha\}\}$. And this can be written as $\bigcup\{\alpha,\{\alpha,\alpha\}\}$. Since we already have an axiom of union, all that we need now is:

Pairing Axiom

If x and y are sets there is a set whose members are exactly x and y (i.e. the doubleton set $\{x,y\}$).

Starting with the null set axiom now, the pairing axiom and the union axiom enable us to construct, inductively, all natural numbers. But what about the "next" ordinal number, ω ? We shall wish to define ω to be the set of all natural numbers. Since we need an axiom in order to do this, we may as well formulate this axiom in the most convenient form, namely:

Axiom of Infinity

There is a set whose elements are all the natural numbers.

It turns out to be enough to assume that some infinite set exists, but the above formulation is adequate. It is this one axiom which takes us from the realm of finite sets into the infinite sets. Once the initial jump from the finite to the infinite has been accomplished, we need make no more raw existence assumptions about sets. (So the axioms of null set and infinity are the only ones which assert that a set simply exists, all other axioms which provide us with new sets being really descriptions of operations which we allow in forming new sets from old ones.)

We now have enough axioms to keep the construction of the ordinals and the Zermelo hierarchy going "for ever". For example, we obtain the second limit ordinal, $\omega + \omega$, by applying the axiom of replacement to the function which sends each n in ω to $\omega + n$, thereby obtaining the set $\{\omega + n \mid n \in \omega\}$, whence we may obtain $\omega + \omega$ as $\omega \cup \{\omega + n \mid n \in \omega\}$.

In fact, we do not need to assume the axiom of subset selection now. This axiom is provable from the axioms of replacement, union, pairing, and null set. To see this, suppose P is a property expressible in LST, and that x is a given set. We wish to prove that $\{y \in x \mid P(y)\}$ is a set. Consider the function F defined on x by

$$F(y) = \begin{cases} \{y\}, & \text{if } P(y) \\ \emptyset, & \text{if } \neg P(y). \end{cases}$$

(Recall that $\{y\} = \{y,y\}$.) Clearly, if P is expressible in LST, so is F. So by the axiom of replacement, $\{F(y) \mid y \in x\}$ is a set. But then by the axiom of union, $\{y \in x \mid P(y)\} = \bigcup\{F(y) \mid y \in x\}$ is a set.

At this point it is tempting to say to oneself "O.K. that does it, we have now written down all the assumptions we need in order to construct a sound and adequate set theory, so we are done." Unfortunately there are two principles which we have overlooked. The first of these has been overlooked because it is so very obvious. If two sets are equal, then clearly they will have the same elements. No problem there, this is a theorem of logic. But we are doing real set theory here, so a set should be determined by what its members are. In other words the converse of the above fact should also be valid: if two sets have exactly the same elements, then they are the same set:

Axiom of Extensionality

Let x and y be sets. If x and y have the same elements, then x = y.

This is the axiom which tells us that we are talking about sets, i.e. collections of objects viewed as objects in their own right.

The last axiom is connected with a point which the reader may have already noticed. Even with our axioms as outlined above, in order to construct the ordinal numbers and the Zermelo hierarchy we must be able to carry out induction arguments: both construction by induction and proof by induction. How can we be sure that this is possible? The answer is, we need another axiom. Comparing our present task of axiomatising set theory with the formulation of the Peano axioms for the natural numbers (a fairly good comparison to make, by the way), the axiom we need here will correspond to the "principle of mathematical induction" of the Peano axioms. What it will say is that the membership relation, $\in$, is well-founded.

Axiom of Foundation

If a is any non-empty set (of sets), there is a member of a which is minimal with respect to the relation $\in$.

Now at last we can sit back and admire our handiwork. Assuming only the eight axioms listed above it is possible to give a rigorous development of set theory, including the construction of the ordinal and cardinal number systems, the Zermelo hierarchy, and indeed the constructible hierarchy. And from this point one is able to define everything one requires for analysis, algebra, etc. Of course, every set theorist would admit that the axioms do not form a particularly attractive collection of principles on which to base mathematics. But none of the axioms can be regarded as at all suspect (can they?). And they were arrived at by taking a fairly natural notion of what the set theoretic universe should look and then analysing what basic assumptions lay behind this picture. So, attractive or not, the phrase "ZF set theory" is nowadays taken to mean the system of set theory whose axioms are the eight axioms listed above.

<u>Appendix II</u> INDEPENDENCE PROOFS IN SET THEORY

The reader will certainly be familiar with the technique of proving a theorem, starting from a set of axioms. But how can we prove that a certain statement is <u>not</u> provable? Indeed, with the theory ZF (or ZFC), could this be even conceivably possible, since everything we do in mathematics is, ultimately, done in ZF (or ZFC)? It was one of the major mathematical achievements of this century when <u>P. J. Cohen</u> found a method whereby, working in ZFC one can prove that certain statements are neither provable nor refutable in the system ZFC. In particular, Cohen demonstrated that the continuum hypothesis could not be proved in ZFC. (Combined with Gödel's earlier result that the CH follows from $V = L$, an axiom which Gödel had shown could not be refutable in ZFC, this established the undecidability of the CH in ZFC. In fact, Gödel's result can be dispensed with, since Cohen's method may also be used to establish the non-provability in ZFC of $\neg$CH.) Since then, Cohen's method has been greatly simplified, and used to demonstrate that many classical open problems in mathematics are, in fact, undecidable in ZFC. We present here a brief outline of this method.

Our starting point is the Zermelo hierarchy. Recall that we obtain $V_{\alpha+1}$ from V_α by setting $V_{\alpha+1} = \mathcal{P}(V_\alpha)$. Now, as everyone knows, associated with every subset of a given set X is a particular function from X to the set $\{0,1\}$, namely the characteristic function of the set, which takes the value 1 on points in the set and the value 0 on points outside the set. Given a subset of X, we "know" its characteristic function, and conversely, given any function from X into $\{0,1\}$ we "know" the set for which it is the characteristic function. Thus, in any set theoretical argument we could work with functions from sets into $\{0,1\}$ instead of with sets. (This would clearly be rather pointless in practice, but presents no great difficulties.) So suppose we define a "Zermelo hierarchy" of characteristic functions as follows:

$$V_0^* = \emptyset ;$$

$$V_{\alpha+1}^* = \{ f \mid f : V_\alpha^* \to \{0,1\} \};$$

$$V_\alpha^* = \bigcup_{\beta<\alpha} V_\beta^* , \text{ if } \alpha \text{ is a limit ordinal.}$$

Then clearly, what we end up with is just a disguised version of the usual Zermelo hierarchy. Indeed, the two hierarchies correspond level by level in a natural way. Thus, if we set

$$V^* = \bigcup_{\alpha\in On} V_\alpha^* ,$$

the class V^* is a functional analogue of the universe, V.

But what is so special about restricting our functions to have only the two values 0 and 1 ? Well, if f is the characteristic function of the set x, then f(a) = 1 means that a is an element of x and f(a) = 0 means that a is not an element of x. And there are no further possibilities. Or are there ? Certainly, from our point of view there are no other eventualities. But are there any internal, mathematical reasons restricting us to just these two. Well, what is the role of the two values 0 and 1 in all of this ? A few moments reflection leads to the conclusion that they are truth values: 1 denotes truth and 0 falsity. The reason why we feel that the functions involved in the definition of V^* should only take the values 0 and 1 is because normal logic only permits two possibilities, true or false. If f_x denotes the characteristis function of x (regarded as a subset of some "big" set U), then for any a (in U), $f_x(a) = 0$ means that the statement "$a \in x$" is false, and $f_x(a) = 1$ means that this statement is true. Let us now see what restrictions are placed on the values of the functions f_x by the logic, ignoring for the moment any restrictions which arise from our intuition. Well, if we have two sets x and y , then for any a, the truth of the statement "$a \in x \cap y$" is equivalent to that of both "$a \in x$" and "$a \in y$", so we must have $f_{x\cap y}(a) = f_x(a).f_y(a)$. Similarly, $f_{x\cup y}(a) = \max(f_x(a),f_y(a))$. Also, $f_{U-x}(a) = 1 - f_x(a)$. Now, although we have so far been regarding the functions as still mapping into $\{0,1\}$, it is possible now to see just what restrictions are placed upon the range space by the logic. It must be a boolean algebra! Although we just allow our characteristic functions to take two values, the mathematics would be just the same if they took values in any boolean algebra. And of course, the set $\{0,1\}$ is a boolean algebra with the operations

$$a.b = a \text{ times } b \;;$$

$$a \vee b = \max(a,b) \;;$$

$$-a = 1 \text{ minus } a.$$

(Which connects in with our remarks above about the functions f_x, of course.) In other words, what is important about the values our "characteristic functions" may take is that they behave in the same manner as truth values (remember truth tables?), and thus reflect the logic involved in the combination of sets and assertions about sets. In fact, when we come to look carefully into this point, we soon realise that what is required is not a boolean algebra but a complete boolean algebra (i.e. a boolean algebra in which every set has a sup and an inf.). This should become clear below.

So let $\mathbb{B}$ be any complete boolean algebra now. We define the $\mathbb{B}$-valued universe of sets as follows:

$$V_0^{(\mathbb{B})} = \emptyset \;;$$

$$V_{\alpha+1}^{(\mathbb{B})} = \{ f \mid f : V_\alpha^{(\mathbb{B})} \longrightarrow \mathbb{B} \} \;;$$

$$V_\alpha^{(\mathbb{B})} = \bigcup_{\beta<\alpha} V_\beta^{(\mathbb{B})} \quad \text{, if } \alpha \text{ is a limit ordinal.}$$

$$V^{(\mathbb{B})} = \bigcup_{\alpha \in On} V_\alpha^{(\mathbb{B})} \;.$$

An element of $V^{(\mathbb{B})}$ will be a $\mathbb{B}$-valued set. But just what is a $\mathbb{B}$-valued set? Well, if f is such an entity, then f provides us with a probability distribution for "the set denoted by f". That is, if $f : V_\alpha^{(\mathbb{B})} \longrightarrow \mathbb{B}$, then for any $u \in V_\alpha^{(\mathbb{B})}$, if $f(u) = 1$, then u is, with probability 1, an element of f in $V^{(\mathbb{B})}$. And if $f(u) = 0$, then u is, with probabliity 0, an element of f, which is the same as saying that with probability 1, u is in the complement of f in $V^{(\mathbb{B})}$. And if $f(u) = b$, then u will be an element of f with probability b in $V^{(\mathbb{B})}$. So in $V^{(\mathbb{B})}$ it is not always the case that a set u is either <u>in</u> a set f or else <u>not in</u> f: it can be <u>partly in</u> and <u>partly out</u>. This state of affairs may strike <u>us</u> as odd, but as far as the internal logic of the system is concerned all is in order. For example, the logic demands that the probability of a set not being in another is the boolean complement of the probability of the first set being in the complement of the second. And this turns out to be the case.

So what do we have so far? We have a "universe of sets". And for the elements of this "universe" we can answer basic membership questions (the answer being always an element of the boolean algebra). But there is more to set theory than saying what the truth value of a statement "$a \in x$" is. For a start, what about statements of set equality? Well, since two sets will be equal iff they have the same members and we can answer membership questions, we can define, in a natural manner, the <u>probability</u>, or <u>truth value</u>, of the statement "$x = y$" for sets x and y in our "universe". We denote by $\| x = y \|$ the probability that $x = y$ holds in our universe. (Although the precise definition of $\| x = y \|$ as an element of the boolean algebra is quite natural, there are some technical complications involved, so we do not give it here.) But this is still not enough. In set theory we want to know the truth value of any assertion about sets. Now, any assertion about sets can be written in the language LST. So what we require here is a definition of the <u>probability</u> or <u>truth value</u>, $\| \varphi \|$, of any sentence φ of LST when interpreted in the boolean valued universe $V^{(\mathbb{B})}$. This is defined by induction on the construction of φ. If φ is a basic membership or equality statement we are done already. And we procced inductively now, setting:

$$\| \varphi \wedge \psi \| = \| \varphi \| \wedge \| \psi \| ;$$

$$\| \varphi \vee \psi \| = \| \varphi \| \vee \| \psi \| ;$$

$$\| \neg \varphi \| = - \| \varphi \| ;$$

$$\| \exists x \varphi(x) \| = \sup_x \| \varphi(x) \| ;$$

$$\| \forall x \varphi(x) \| = \inf_x \| \varphi(x) \| .$$

(We may obtain the definition for $\varphi \to \psi$ from the identity

$$\varphi \to \psi \equiv \neg \varphi \vee \psi ,$$

and that for $\varphi \leftrightarrow \psi$ from

$$\varphi \leftrightarrow \psi \equiv (\varphi \rightarrow \psi) \wedge (\psi \rightarrow \varphi) \text{ .)}$$

Notice that there is some duplication of notation here. All symbols to the left of the equality signs are logical symbols, and all those to the right are boolean operations in $\mathbb{B}$. But there is no need to try to avoid this. Indeed, it emphasises the reason why a complete boolean algebra is required. The operations of a boolean algebra behave exactly as do their canonical logical counterparts.

At this point let us warn the reader that we have vastly simplified matters in the above account, and that he should not concentrate on any details. Our aim is just to outline the method, not to present it. Hence absolute accuracy has been sacrificed for simplicity. (But not to too great an extent, of course!) As to the next two, highly crucial parts of the development, we shall not attempt to indicate how the various proofs go: they are highly technical.

It turns out that all of the axioms of ZFC are valid in $V^{(\mathbb{B})}$ (for any $\mathbb{B}$) with probability 1. Moreover, if we can deduce a statement ψ from a statement φ by means of a logical argument, then $\|\varphi\| \leq \|\psi\|$. Hence any theorem of ZFC will be valid in $V^{(\mathbb{B})}$ with probability 1. And, of course, it follows that the negation of any theorem of ZFC will be valid with probability 0.

Suppose now that we are given an assertion φ about sets (e.g. CH), and we wish to demonstrate that φ is not decidable in the system ZFC. We assume that φ has, in fact, not been decided! Which is the same as saying that we have been unable to calculate $\|\varphi\|$ in the universe $V^{(\mathbb{B})}$ when $\mathbb{B}$ is the particular algebra $\{0,1\}$. But perhaps, by a careful choice of algebra $\mathbb{B}$, we can prove that in $V^{(\mathbb{B})}$, $\|\varphi\|$ lies strictly between 0 and 1. Then we shall know at once that neither φ nor its negation is provable in ZFC. And there is our method!

As might be expected, the probability distributions and the truth values of statements about sets in a boolean valued universe depend heavily upon the actual boolean algebra involved. Hence, in order to prove the independence (in ZFC) of a particular statement, the first step is to find a suitable boolean algebra . This can be extremely difficult. Indeed, most results require the explicit construction of a boolean algebra. The "standard examples" rarely suffice. Then there is the problem of proving that the statement in question has a truth value strictly between 0 and 1 in the "universe" chosen. This can also prove tricky. Indeed, on occasions what one ends up doing is finding <u>two</u> boolean algebras, and proving that the statement in question is true with probability 1 in one universe and with probability 0 in the other! (This also does the trick, of course.)

Thus, although one is ultimately proving that some statement is not provable, what one actually does when applying the method outlined above is just construct a boolean algebra and prove a lemma about that algebra. All this can be done in good old ZFC. (Although, by virtue of our discussion on constructibility, it would be permissible to use V = L as well, and indeed this is sometimes necessary.) So even when he is proving some independence result, a set theorist is just doing <u>set theory</u>!

GLOSSARY OF KEY TERMS

SPECIAL SYMBOLS

SYMBOL	APPROX. MEANING (where possible)	PAGE
c.c.c.	countable chain condition	2
$\lhd$	subgroup	2
G/H	factor group	2
$+$	group operation	
$\oplus$	direct sum	
$\dot{\oplus}$	external direct sum	
$\langle A \rangle$	subgroup generated by A	
$\underset{\sim}{Z}$	group of integers	2
$G \xrightarrow{\varphi} H$	group homomorphism	
$\mathrm{Im}(\varphi)$	image of φ	
$\mathrm{Ker}(\varphi)$	kernel of φ	
$\mathrm{Hom}(G,H)$	group of homomorphisms from G to H	
1_X	identity on the set/group X	3
$f:X \longrightarrow Y$	function from X to Y	
$f[A]$	image of f on A	
$f^{-1}[A]$	pre-image of f on A	
α, β, γ	ordinal numbers	7
$\alpha+1$	first ordinal after α	7
$\lvert X \rvert$	cardinality of X	8
ω	{first infinite ordinal {set of all natural numbers	8
κ, λ	infinite cardinal numbers	8
ω_α	α'th uncountable cardinal as an ordinal	9
$\aleph_0$	ω as a cardinal	9
$\aleph_\alpha$	α'th uncountable cardinal as a cardinal	9
κ^+	first cardinal after κ	9
(x,y)	ordered pair of x and y	12

SUGGESTED FURTHER READING

For short accounts of set theory, formal languages, model theory, and other parts of mathematical logic, see the various articles in

[Ba] K. J. Barwise (Ed): Handbook of Mathematical Logic , North Holland (1977)

For a reasonably detailed development of constructibility theory, see

[De] K. J. Devlin: Constructibility , in [Ba] above.

For the reader who wants to fill in the gap we left in our solution of the Whitehead Problem, the only source at present is Shelah's original paper:

[Sh] S. Shelah: A Compactness Theorem for Singular Cardinals, Free Algebras, Whitehead Problem and Transversals, Israel Journal of Mathematics 21 (1975), 319 - 349.

As a matter of historical interest, let us also mention the original monograph on constructibility:

[Gö] K. Gödel: The Consistency of the Axiom of Choice and of the Generalised Continuum Hypothesis. Annals of Mathematics Studies 3, Princeton University Press (1940).